AAPG Continuing Education Course Note Series #12

SECONDARY RESERVOIR POROSITY IN THE COURSE OF SANDSTONE DIAGENESIS

by

VOLKMAR SCHMIDT

&

DAVID A. McDONALD

*PETRO-CANADA, GEOLOGICAL RESEARCH,
CALGARY, ALBERTA*

A complete slide and cassette tape presentation of this course can be purchased from Science-Thru-Media Co., 295 Fifth Avenue, Room 819, New York, New York 10016

Offered by the
AAPG Education Department

Other titles in AAPG Continuing Education Course Note Series:

#1 **Plate Tectonics and Hydrocarbon Accumulation.**
William R. Dickinson and Hunter Yarborough, 83 pages, $5.00.

#2 **Deltaic and Shallow Marine Sandstones: Sedimentation, Tectonics and Petroleum Occurrences**
Robert J. Weimer, 169 pages. $6.00.

#3 **Exploration for Sandstone Stratigraphic Traps**
Robert R. Berg, 35 pages. $3.00.

#4 **Pore Pressure: Fundamentals, General Ramifications and Implications for Structural Geology**
Peter E. Gretener, 87 pages. $5.00.

#5 **Geology of Continental Margins**
Hunter Yarborough, Kenneth O. Emery, William R. Dickinson, Donald R. Seely, Wallace G. Dow, Joseph R. Curray, and Peter R. Vail, 122 pages. $6.00.

#6 **Deep-Water Oil Sand Reservoirs — Ancient Case Histories and Modern Concepts**
Oscar E. Weser, 170 pages with color map. $6.00.

#7 **Deposition, Diagenesis, and Hydrocarbon Potential of "Deeper-Water" Limestones**
Peter A. Scholle, 27 pages. $2.00.

#8 **Physical and Chemical Constraints on Petroleum Migration**
Colin Barker, Henry H. Hinch, Robert W. Jones, Clayton McAuliffe, James Momper, and Leigh Price, 224 pages. $8.00.

#9 **Concepts and Methods of Subsurface Facies Analysis**
Richard C. Selley, 88 pages. $5.00.

#10 **Organic Geochemistry In Petroleum Exploration**
Colin Barker, 159 pages. $8.00.

#11 **Geology of Carbonate Porosity**
Don Bebout, Graham Davies, Clyde H. Moore, Peter A. Scholle and Norman C. Wardlaw, 248 pages. $10.00.

#12 **Secondary Reservoir Porosity in the Course of Sandstone Diagenesis**
Volkmar Schmidt and David A. McDonald, 125 pages. $7.00.

These are author-prepared publications distributed by the
Education Department
American Association of Petroleum Geologists
P.O. Box 979
Tulsa, OK 74101

TABLE OF CONTENTS

SUMMARY

Secondary porosity in sandstones can be classified according to origin and pore texture. Five significant genetic classes of secondary porosity are defined by the following processes of origin: (1) fracturing; (2) shrinkage; (3) dissolution of sedimentary grains and matrix; (4) dissolution of authigenic pore filling cement; and, (5) dissolution of authigenic replacive minerals. Hybrid pores are characterized either by the coexistence of several genetic classes of secondary porosity or by the coexistence of primary and secondary porosity.

Secondary porosity appears in five major groups of pore textures: (1) intergranular pores; (2) oversized pores; (3) moldic pores; (4) intra-constituent pores; and, (5) open fractures. Some secondary porosity mimics the entire range of pore sizes and pore textures of primary sandstone porosity. Other secondary porosity bears a general resemblance to the textures of primary porosity but differs in detail. Secondary porosity may also appear in textures that are entirely different from those of primary porosity.

In most instances it is possible to identify the occurrence of secondary porosity in thin section using a set of simple petrographic criteria that include: (1) partial dissolution; (2) molds; (3) inhomogeneity of packing; (4) oversized pores; (5) elongate pores; (6) corroded grains; (7) intra-constituent pores; and, (8) fractured grains. In medium or coarse-grained sandstones secondary porosity can, in some instances, be observed by the naked eye.

The detailed analysis of the petrological attributes of

secondary porosity may require the use of advanced analytical techniques, such as cathode luminescence petrography, scanning electron microscopy, pore cast examination, microprobe analysis and stable isotope analysis. In most sandstones, however, a surprising amount of important information can be obtained simply by the careful and methodical use of conventional petrographic microscopy.

Secondary porosity plays an important role in the diagenesis of sandstones. The volume of secondary porosity equals or exceeds that of primary porosity in the sandstones of many sedimentary basins worldwide. A significant percentage of the world's reserves of natural gas and crude oil are contained in secondary sandstone porosity. Prudhoe Bay Field and the Jurassic fields of the North Sea are examples of the many giant hydrocarbon accumulations in secondary sandstone porosity.

Chemical, physical, physicochemical, biochemical and biophysical processes result in secondary sandstone porosity through leaching and shrinkage of rock constituents, or through the opening of fractures and porous burrows and borings. Secondary sandstone porosity can originate anywhere in the sedimentary crust: (1) before effective burial under the direct physico-chemical influence of the environment of deposition (eogenetic)[1]; (2) during effective burial at any depth above the zone of metamorphism (mesogenetic)[1]; and (3) during exposure following a period of burial (telogenetic)[1]. Secondary porosity occurs in sandstones of any mineralogical or textural composition and of any Phanerozoic age. It is most common in sandstones that have undergone relatively long

---

[1]Diagenesis terminology adopted from Choquette & Pray, 1970

lasting and deep burial and have lost their primary porosity.

Leaching of rock constituents creates by far the largest percentage of secondary sandstone porosity. The list of dissolved rock constituents includes the majority of minerals found in clastic sediments. However, most of the secondary porosity in ancient sandstones originated as a result of mesogenetic leaching of the carbonate minerals calcite, dolomite and siderite. This decarbonatization removes depositional carbonate constituents and diagenetic carbonate such as cements or replacements.

Mesogenetic decarbonatization of sandstones can be caused by a number of chemical and physico-chemical diagenetic processes. However, it appears to be related mainly to the decarboxylation of organic matter in strata adjacent to the sandstones during the course of organic maturation. The process of decarboxylation leads to the generation of carbon dioxide which, in the presence of water, produces carbonic acid. It is suggested that such acidic waters come in contact with sandstones and react with their carbonate constituents.

In most instances it is possible to differentiate microscopically between primary and secondary sandstone porosity. Thus it is possible to trace the loss of primary porosity during burial. In the presence of water and hydrostatic pressure, primary sandstone porosity cannot exist beyond specific limits of temperature-time exposure except for a small volume of irreducible lamellar porosity between grains. The limiting temperature-time exposure increases with increasing mineralogical stability of the sandstones and, subordinately, with increasing grain size.

The mesodiagenesis of sandstones can be divided into four

stages: (1) immature mesodiagenesis - mechanical compaction; (2) semi-mature - chemical compaction of primary porosity; (3) mature - only secondary porosity present; and, (4) supermature - no effective primary or secondary porosity.

Decarbonatization may create considerable quantities of secondary porosity during semi-mature mesodiagenesis. However, the average addition of carbonate to the sandstone in this diagenetic stage usually exceeds the average carbonate removal.

Decarbonatization culminates during mature mesodiagenesis at which stage it greatly outweighs carbonatization. Much secondary sandstone porosity, therefore, originates after effective primary porosity has been lost. Fractures and irreducible lamellar porosity apparently provide sufficient access for decarbonatizing fluids to start the leaching process even in sandstones of low permeability.

Enormous volumes of carbonate move upwards in solution from diagenetically mature sandstones and are, at least in part, reprecipitated in immature and semi-mature sandstones. Within a subsiding prism of clastic sediments much of the carbonate content is being recycled upwards in this fashion and sandstones at shallower depths are being enriched in carbonate.

Primary migration of hydrocarbons commonly follows closely after the secondary porosity has been formed. This is apparently related to the fact that in the maturation of organic matter, the main phase of hydrocarbon generation follows after the culmination of decarboxylation. This close association of source and reservoir in time and space favours the accumulation of hydrocarbons in secondary porosity.

In the presence of pore water, secondary porosity is

gradually reduced during deep burial although at a much slower rate compared to primary porosity.

The main geological and economic significance of secondary sandstone porosity is that it extends the depth range for effective sandstone porosity far below the depth limit for effective primary porosity. Generation and primary migration of hydrocarbons occurs mainly below the range of effective primary porosity. The path of primary migration and the site of accumulation of hydrocarbons are, therefore, commonly controlled by the distribution of secondary porosity.

The geometry of secondary porosity resembles that of primary porosity only in those cases where the nonsoluble framework of the sandstone remained unchanged after cementation and where the pore filling cement was later exclusively and completely dissolved. In most instances, however, secondary porosity differs from primary porosity in comparable sandstones in the following geometric aspects: (1) the range of pore sizes and pore shapes is wider; (2) the spectra of sizes and shapes of pore throats are radically different; (3) the distribution of porosity within individual bodies of sandstone is more variable. As a result of their pore geometry, most sandstones with secondary porosity show strikingly different reservoir characteristics and physical properties if compared with sandstones of similar original lithology that show a comparable percentage of primary porosity. The affected parameters include: (1) permeability; (2) irreducible water saturation; (3) mechanical competence; (4) sonic travel time. The presence of secondary porosity commonly modifies the response to several well-logging methods. It also should noticeably affect geophysical characteristics. Relatively low water saturations are characteristic for many hydrocarbon reservoirs in sandstones with secondary porosity.

The reservoir aspects of fields producing from secondary sandstone porosity are often more like those of carbonate reservoirs rather than those of fields with primary sandstone porosity. The distribution of secondary porosity in a prospective or productive sandstone formation may not necessarily show a direct relationship with depositional lithofacies or burial history and can be difficult or even impossible to predict with conventional subsurface methods. However, detailed geological and petrological analysis of the factors that control the occurrence of secondary porosity often greatly enhance the understanding and prediction of its distribution. Sandstone reservoirs with secondary porosity have special problems but they also offer special opportunities for both the explorationist and the exploitationist. It is necessary, firstly, to recognize the secondary nature of the porosity of these reservoirs; secondly, to interpret controls and timing of the origin of their porosity; thirdly, to evaluate any factors that may have played a role in preserving this porosity; and, fourthly to survey the existing pore geometry in order to optimize the methods and economics of finding and recovering.

INTRODUCTION

The subject of secondary porosity in sandstones received little attention in the published literature prior to 1975. Most workers considered only a very small portion of sandstone porosity to be of secondary origin. Except for fractures, most accounts of secondary porosity described outcropping sandstone formations where invading meteoric water had removed soluble constituents (Sedimentation Seminar, 1969; Hraber & Potter, 1969). Krynine (1941) indicated that leaching of carbonate in the Oriskany Sandstone may possibly have occurred in the subsurface. Proshlyakov (1960) was probably the first author to convincingly report significant amounts of sandstone porosity that formed in the subsurface in the presence of saline formation water. The first detailed discussion of the possible processes which create secondary porosity in the subsurface was published by Savkevic (1971). Rowsell and De Swardt (1973) observed secondary porosity of probable subsurface origin in the Middle Ecca sandstones of Pennsylvanian age in the northern Karoo basin of South Africa. They were the first authors outside Russia to apply the concepts developed by Russian geologists concerning mesogenetic porosity. Until 1976, these Russian and South African studies went largely unnoticed in North America. Unaware of the above studies, several groups of North American geologists have, during the last several years recognized the common occurrence of secondary sandstone porosity that originated in the subsurface.

Recently, results of these investigations have appeared in the literature and several workers have described occurrences from the subsurface where the porosity originated after burial (Schmidt, 1976; Stanton & McBride, 1976; Alcock & Benteau, 1976; Pittman, 1979; Hayes, 1979). Hayes and the authors of the present paper recognized the widespread occurrence of secondary sandstone porosity as early as 1970. Subsequently the authors refined the criteria for recognizing secondary porosity (Schmidt et al, 1977) and studied the phenomenon of secondary sandstone porosity in various geological settings in Canada, Alaska and Europe. A study of the Parsons Sandstone of the Mackenzie Delta (Schmidt, ibid.) provided several new concepts on the origin of secondary porosity in the subsurface. McBride and his students have applied these concepts and criteria very successfully when investigating secondary porosity in the subsurface of southern Texas (Stanton & McBride, 1976; Lindquist, 1976).

As the authors' studies progressed, it became apparent

that much secondary porosity has previously been misidentified as primary porosity, and their views on the abundance of secondary porosity have been changing with increasing information (Figure 1). At the present time it is certain that at least one-third of total sandstone porosity is of secondary origin. It appears possible that secondary porosity in sandstone is more abundant than primary sandstone porosity. Undoubtedly, secondary sandstone porosity is of major geological significance. Its origin, stability, and distribution form important aspects of sandstone diagenesis.

At present secondary sandstone porosity, with the exception of fractures, is still largely mistaken for primary porosity and commonly goes unnoticed. This is due to the wide variety of textures of secondary porosity which often mimic those of primary porosity, and to the fact that most investigators are unfamiliar with the diagnostic criteria for recognizing secondary porosity in sandstones.

Secondary porosity constitutes the predominant or exclusive form of effective porosity in the reservoir sandstones of many oil and gas fields. The exploration and exploitation of such sandstone reservoirs will clearly benefit from an accurate identification of the secondary nature of the porosity and of its textural relationship with the host rock (Hayes, 1979; Pittman, 1979).

In general, the study of secondary sandstone porosity is still very much in its infancy. However, the available information is sufficient to bring the emerging role of secondary porosity in sandstone diagenesis into perspective.

The intentions of this Short Course are threefold:

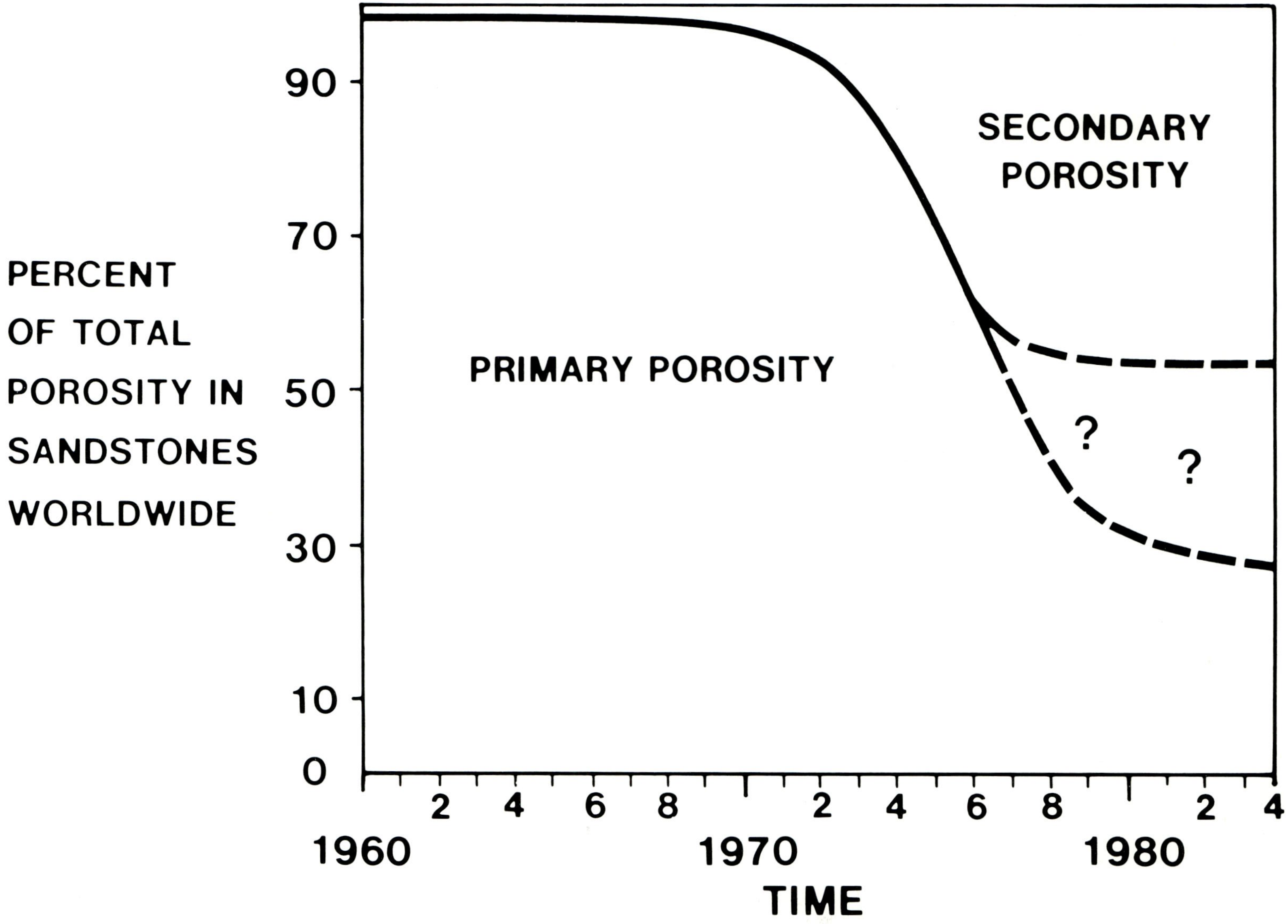

FIG. 1 - Change of interpretation of nature of sandstone porosity.

(1) to outline the textural spectrum of secondary sandstone porosity and to provide criteria which facilitate its recognition;

(2) to provide a preliminary account of the occurrence of secondary porosity in sandstones and to follow its evolution through the course of diagenesis;

(3) to develop some insight into the reservoir aspects of secondary sandstone porosity.

In this way, we hope, useful stepping stones will be established for those who pursue the fascinating subject of sandstone diagenesis and apply their talents to unlock the resources contained in sandstones with secondary porosity. Specifically we hope that secondary porosity in sandstones can be more consistently reported, its origin more accurately interpreted and its reservoir characteristics more profitably utilized.

These course notes are essentially a compilation of three publications by the authors (Schmidt et al, 1977; Schmidt and McDonald, 1979/a; Schmidt and McDonald, 1979/b). However, it was impossible to include colored microphotographs in this account. The reader is, therefore, referred to the publications cited before and to the excellent microphotographic atlas of sandstone porosity and diagenesis (Scholle 1979).

GENETIC-TEXTURAL CLASSES OF SECONDARY SANDSTONE POROSITY

Secondary porosity in sandstones is mainly the result of dissolution of non-silicate constituents, predominantly carbonate minerals. The soluble precursors of this dissolution porosity occur in three textural forms: sedimentary material, authigenic cement, and authigenic replacement. Fracturing and shrinkage of rock constituents may also create significant secondary porosity in sandstones, though usually in subordinate amounts.

It is useful, for geological interpretations, to base the definition of the principal classes of secondary porosity on textural origin. In addition, the textures in which secondary porosity occurs must be classified in such a manner as to make their description easy and objective.

Principal genetic classes. - We recognize five principal classes of secondary porosity that can be differentiated on the basis of process of origin and textural relationships (Figure 2):

(1) porosity created by fracturing;
(2) porosity created by shrinkage;
(3) porosity created by dissolution of sedimentary constituents;
(4) porosity created by dissolution of authigenic cementing minerals;
(5) porosity created by dissolution of authigenic replacive minerals.

Porosity created by fracturing encompasses any newly formed fractures (Figure 3) including those formed due to stresses resulting from shrinkage of rock constituents or whole rocks. If fractures become filled by cement or sediment then these fillers or their replacive successors may be selectively dissolved giving rise to second cycle fracture porosity. Such reopened fractures would be included in the appropriate class of dissolution porosity.

Shrinkage porosity, forms through dehydration and/or recrystallization of a number of minerals such as glauconite

FIG. 2 - Genetic classes of secondary sandstone porosity.

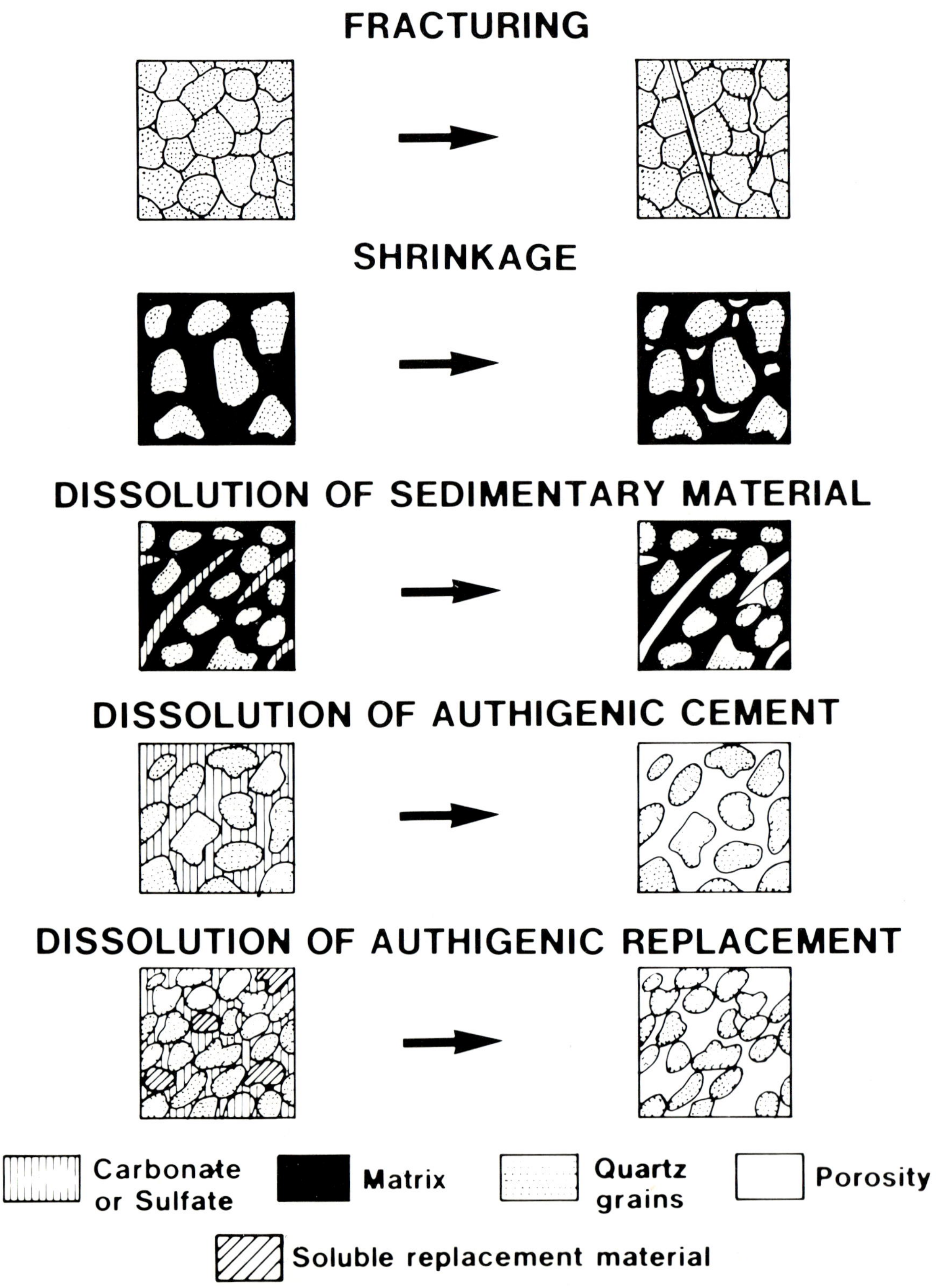

FIGS. 3 to 5 - Examples of genetic classes of secondary sandstone porosity

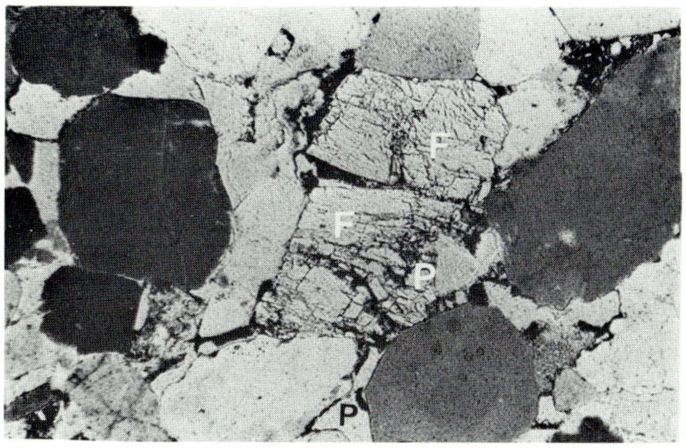

FIG. 3 - Examples of fracture porosity:

secondary pores resulting from fracturing of feldspar grains. Half crossed nicols. Lower Cretaceous, Mackenzie Delta, Canada. P = porosity; F = fractured feldspar.

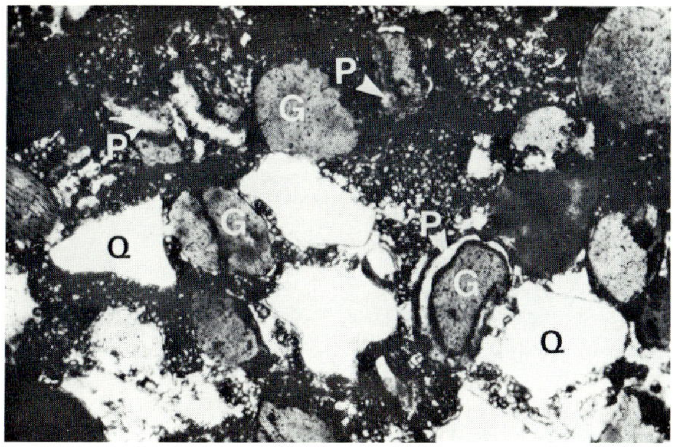

FIG. 4 - Example of shrinkage porosity:

secondary pores resulting from shrinkage of glauconite grains. Polarizer only. Cretaceous, Scotian Shelf, Canada. P = porosity; G = glauconite; Q = quartz.

FIG. 5 - Example of porosity after dissolved sedimentary material:

secondary pores resulting from dissolution of sedimentary siderite matrix and siderite grains. Polarizer only. Jurassic, Statfjord field, North Sea. P = porosity; S = microcrystalline siderite.

(Figure 4) and hematite, or aggregates such as mud. It excludes shrinkage features. Shrinkable constituents may occur as grains, parts of grains, matrix, authigenic cement and authigenic replacement. Accordingly, shrinkage porosity consists of a large variety of porosity textures. Shrinkage pores generally vary in size from submicroscopic dimensions to the dimensions of adjacent sand grains. However, larger shrinkage pores may occur.

Porosity created by dissolution of sedimentary material is very common and important. It results from the selective dissolution of soluble grains and soluble matrix. These are in most cases composed of carbonate minerals (Figure 5). A large number of porosity textures arise from the dissolution of sedimentary constituents. Pore size varies from submicroscopic voids a fraction of a micron in diameter to vugs several centimeters across. Voids ιrequently exceed the size of adjacent grains.

Porosity originating from dissolution of authigenic cement is probably the most common porosity class. The majority of dissolved cements consist of the carbonate minerals calcite, dolomite and siderite (Figure 6). The precursor cements may have occupied any type of primary or secondary porosity. Dissolution of cement may, therefore, reopen any conceivable type of primary and secondary porosity textures.

Porosity resulting from dissolution of authigenic replacive minerals commonly forms a significant percentage of secondary sandstone porosity. It originates from selective dissolution of soluble minerals, dominantly calcite, dolomite and siderite, that previously replaced sedimentary constituents and/or authigenic cements, (Figure 7). The precursor replacement itself may have been either "selective" or "non-

FIGS. 6 to 8 - Examples of genetic classes of secondary sandstone porosity

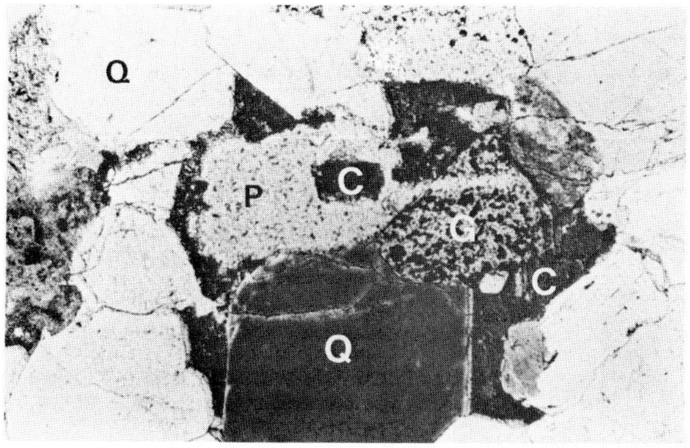

FIG. 6 - Example of porosity after dissolved cement: secondary pore resulting from dissolution of intergranular cement. Half crossed nicols. Lower Cretaceous, Mackenzie Delta, Canada. P = porosity; C = calcite; Q = quartz grains; G = glauconite grain.

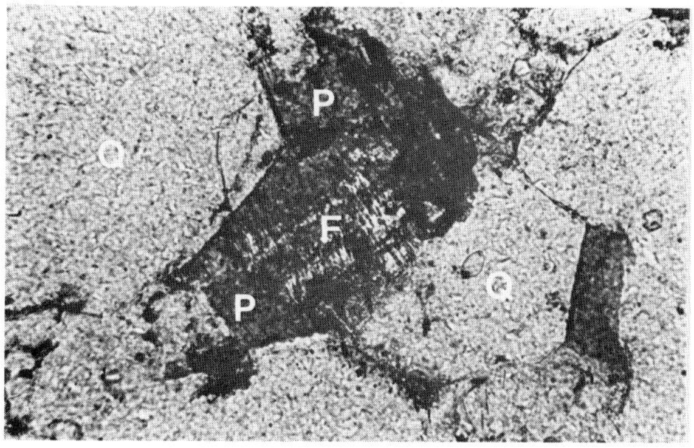

FIG. 7 - Example of porosity after dissolved replacement: secondary pores resulting from dissolution of replacive siderite that partially replaced feldspar grain. Polarizer only. Lower Cretaceous, Mackenzie Delta, Canada. P = porosity; F = feldspar (honeycombed grain); Q = quartz.

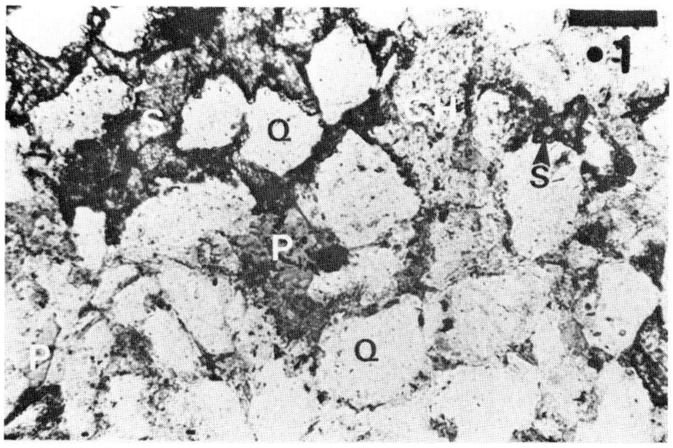

FIG. 8 - Example of hybrid porosity: hybrid secondary pores resulting from the dissolution of intergranular siderite cement and siderite that replaced the margins of quartz and chert grains. Polarizer only. Triassic, Prudhoe Bay Field, North Slope, Alaska. P = porosity; S = siderite (high relief); Q = quartz grains; CH = chert grains (low relief).

selective" as defined by Choquette & Pray (1970). Dissolution of replacive minerals, as with the dissolution of cements, may create any imaginable type of porosity texture.

Pores of hybrid origin. - Individual pores may be of complex parentage either being composed of several genetic classes of secondary porosity or being part primary and part secondary in origin. Such pores of hybrid origin can be termed "hybrid pores" (Figure 9). For example, replacement of the margins of sand grains frequently occurred simultaneously with the cementation of the adjacent intergranular spaces by the same mineral, (Figure 8). Complete dissolution of the authigenic mineral will create hybrid pores because both pore cement and grain replacement are removed.

A large percentage of secondary porosity is found in hybrid pores and they may appear in the form of all the pore textures shown by secondary porosity. Generally, individual pores of each porosity texture can form through numerous combinations of genetic classes.

TEXTURAL SPECTRUM OF SECONDARY SANDSTONE POROSITY

The textures in which secondary sandstone porosity occur must be classified in such a manner as to make their description objective, informative and manageable. Several descriptive groups of pore textures can be differentiated in sandstones using, in part, the nomenclature for pore textures that Choquette & Pray (1970) proposed for carbonate rocks. These groups include the following:

(1) intergranular pore textures;
(2) oversized pore texture;
(3) moldic pore textures;

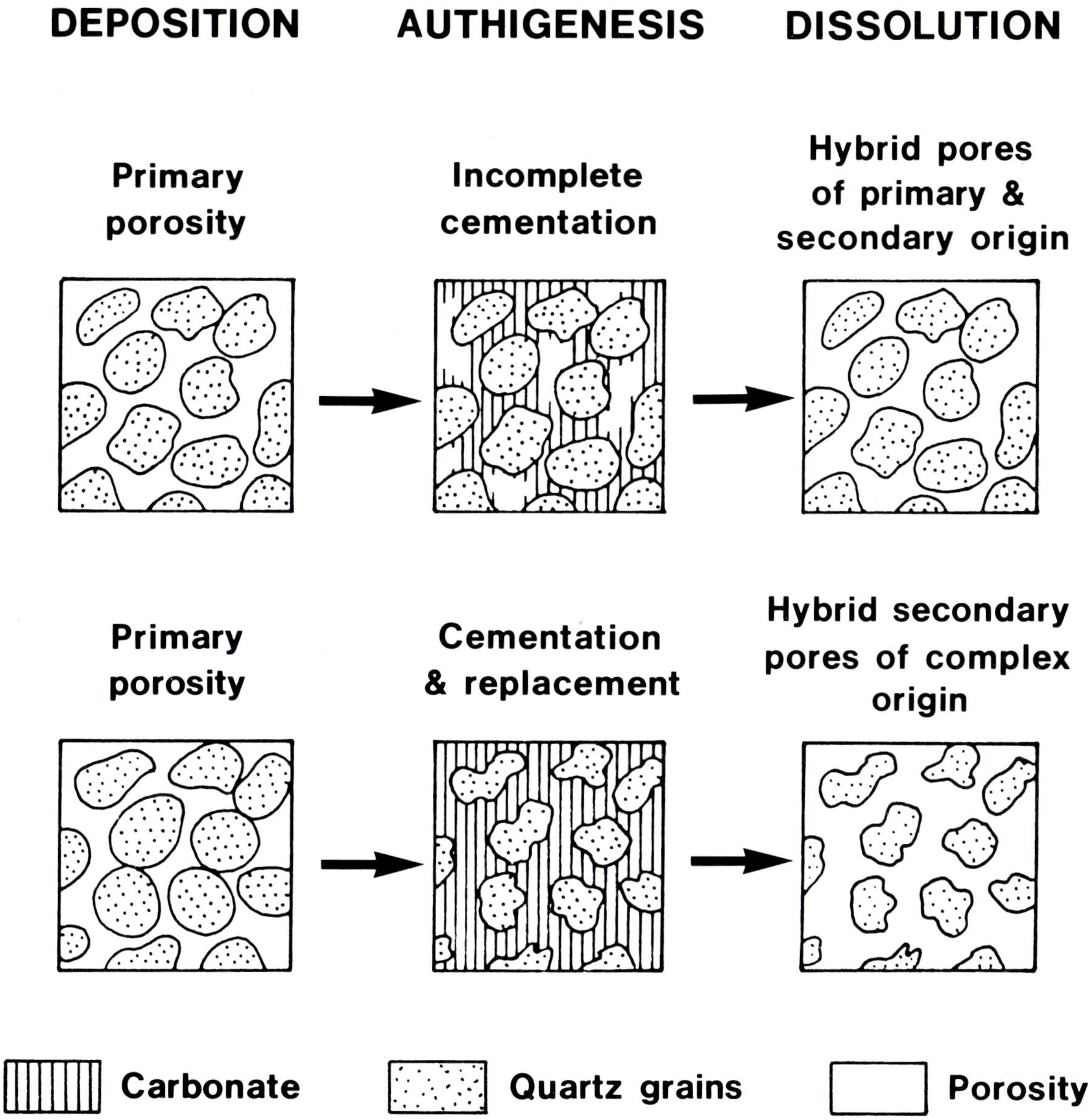

FIG. 9 - Textural development of hybrid pores.

(4) intra-constituent pore textures;

(5) fracture pore textures.

Each group of pore textures consists of two or more distinct types of pore textures. The textural spectrum of secondary sandstone porosity is represented by these types of pore textures as they occur in various genetic classes of secondary porosity (Table 1).

Intergranular pore textures. - Pores between grains form the textural group of intergranular pores (Figure 10). The intergranular pores may be lined by fringing cement or syntaxial cement. Three types of secondary intergranular pore textures can be distinguished: (1) regular intergranular; (2) reduced intergranular; and (3) enlarged intergranular.

Regular intergranular pore texture essentially reflects depositional intergranular spaces in size and shape (Figures 11 & 12), and is identical with primary intergranular pore texture. The secondary pores may occupy intergranular spaces entirely or they may share individual intergranular spaces with matrix, cement, replacement or primary porosity thus forming "complete" or "partial" secondary pores of this texture.

Secondary pores of regular intergranular texture form a significant percentage of secondary sandstone porosity. They result from the following: (1) shrinkage of intergranular sedimentary matrix (e.g. shrinkage of glauconite matrix); (2) dissolution of intergranular sedimentary matrix (e.g. dissolution of carbonate matrix); (3) dissolution of intergranular cement (e.g. dissolution of intergranular carbonate and sulfate cement); (4) dissolution of intergranular replacement (e.g. dissolution of carbonate that replaced glauconite matrix); and, (5) any combination of these genetic-

TABLE 1

TEXTURAL SPECTRUM OF SECONDARY SANDSTONE POROSITY

POROSITY TEXTURES	GENETIC CLASSES OF SECONDARY POROSITY				
	RESULT OF FRACT-URING	RESULT OF SHRINK-AGE	RESULT OF DIS-SOLUT-ION OF SEDIM-ENT	RESULT OF DIS-SOLUT-ION OF CEMENT	RESULT OF DIS-SOLUT-ION OF REPLACE-MENT
INTERGRANULAR TEXTURES:					
REGULAR INTERGRANULAR		X P	X P&C	X P&C	X P&C
REDUCED INTERGRANULAR		X P	X P&C	X P&C	X P&C
ENLARGED INTERGRANULAR		X P	X P&C	X P&C	X P&C
OVERSIZED TEXTURES:					
OVERSIZED FABRIC SELECTIVE		X	X	X	X
OVERSIZED CROSSCUTTING				X	X
MOLDIC TEXTURES:					
GRAIN MOLD		X P	X P&C	X P&C	X P&C
CEMENT MOLD		X P		X P&C	X P&C
REPLACEMENT MOLD		X P		X P&C	X P&C
INTRA-CONSTITUENT TEXT.:					
INTRAGRANULAR		X	X	X	X
INTRA-MATRIX		X	X	X	X
INTRA-CEMENT		X		X	X
INTRA-REPLACEMENT		X	X	X	X
FRACTURE TEXTURES:					
ROCK FRACTURES	X		X P&C	X P&C	X P&C
GRAIN FRACTURES	X			X P&C	X P&C
INTERGRANULAR FRACTURES	X			X P&C	X P&C

P&C indicates open void may extend over part of the textural precursor or over the complete textural precursor.

P indicates open void may extend only over part of textural precursor.

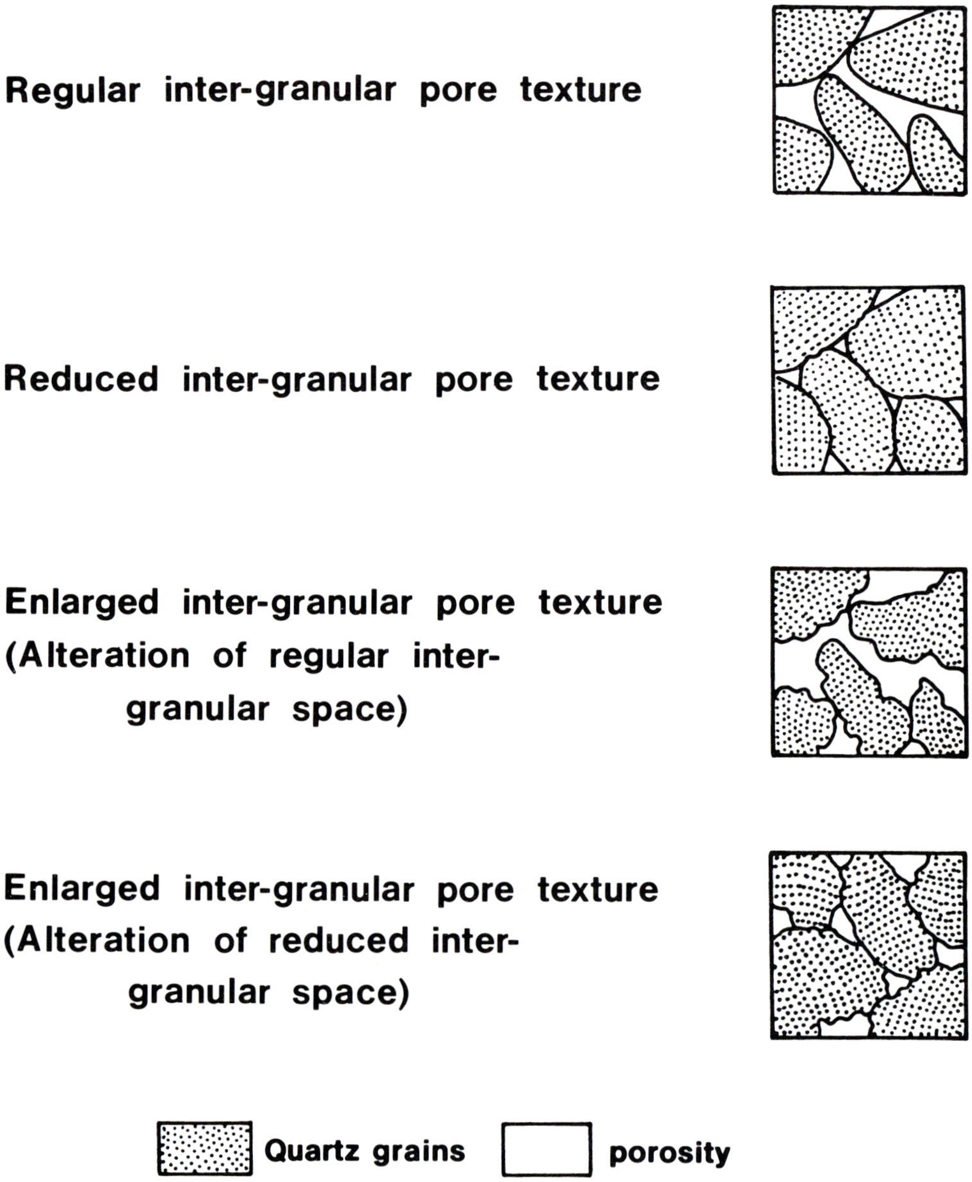

FIG. 10 - Intergranular textures of secondary porosity

FIGS. 11 to 13 - Examples of secondary intergranular pore textures in sandstones

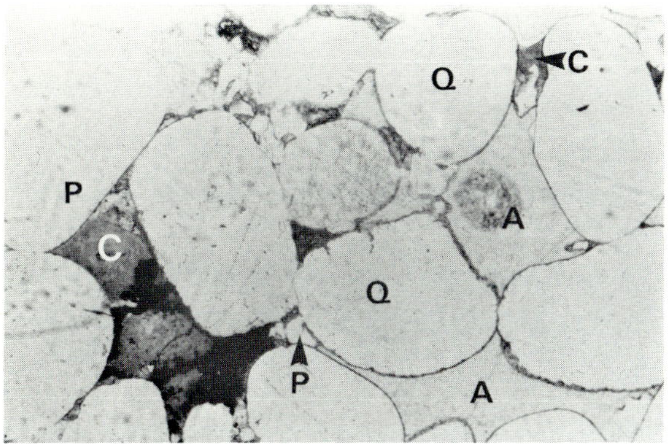

FIG. 11 - Regular intergranular pores of primary origin filled by calcite and anhydrite cement. Incipient dissolution of cement created partial secondary pores or regular intergranular texture. Polarizer only. Cambrian, Northwest Territories, Canada. Union I.O.E. Stopover L-34, 2794 ft (852 m). P = porosity; Q = quartz grains; C = calcite cement; A = anhydrite cement.

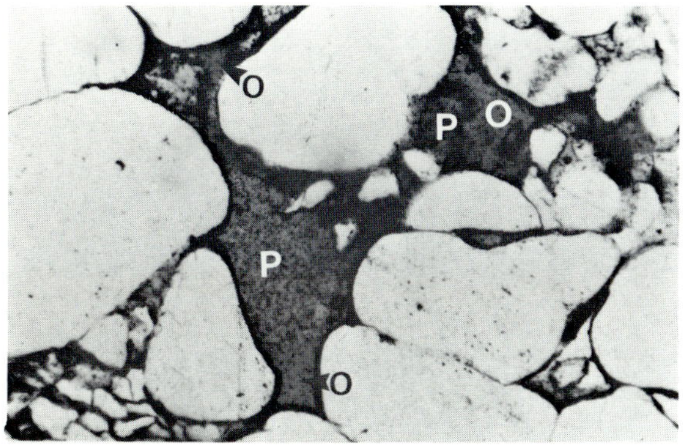

FIG. 12 - Regular intergranular pores of secondary origin resulting from dissolution of intergranular carbonate and sulfate cement. Polarizer only. Same location as for Fig. 11. Depth 2786 ft. (849 m). P = porosity; Q = quartz grains; O = residual oil.

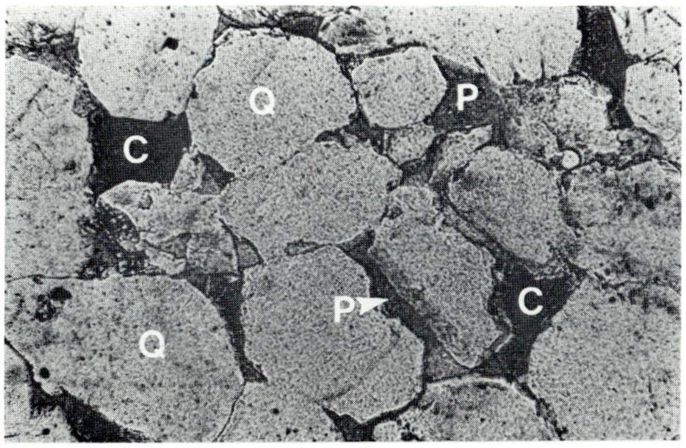

FIG. 13 - Secondary pores of reduced intergranular texture resulting from dissolution of intergranular calcite cement. Polarizer only. Cambrian, Northwest Territories, Canada. P = porosity; Q = quartz; C = calcite.

textural processes of origin (e.g. dissolution of both carbonate cement and carbonate matrix within the same intergranular space).

Reduced intergranular pore texture develops where depositional intergranular spaces are reduced by any of the following: (1) grain deformation; (2) chemical compaction through dissolution at grain contacts; (3) syntaxial cementation; and, (4) fringing cementation. This pore texture mimics that of reduced primary intergranular porosity (Figure 13).

Secondary pores of reduced intergranular texture represent a considerable portion of secondary sandstone porosity. They occur both in the partial and in the complete form. The genetic porosity classes and hybrid genetic pore types encountered are the same as those listed for regular intergranular pore texture. The secondary porosity appearing in the reduced intergranular texture commonly resists further modification (Figure 14). However, regular intergranular pores of secondary origin and enlarged intergranular pores can be more readily altered to reduced intergranular pores during further diagenesis.

Enlarged intergranular pore texture forms either by enlargement of depositional intergranular space or by enlargement of previously reduced intergranular space (Figure 15). This enlargement of the intergranular space can take place in three ways: (1) shrinkage of the grain framework (e.g. shrinkage of glauconite grains); (2) dissolution of grain margins, syntaxial cement or fringing cement (e.g. dissolution of the margins of siderite grains); and, (3) replacement of grain margins, syntaxial cement or fringing cement followed by selective dissolution of the replaced mineral (e.g. marginal replacement of quartz grains by siderite

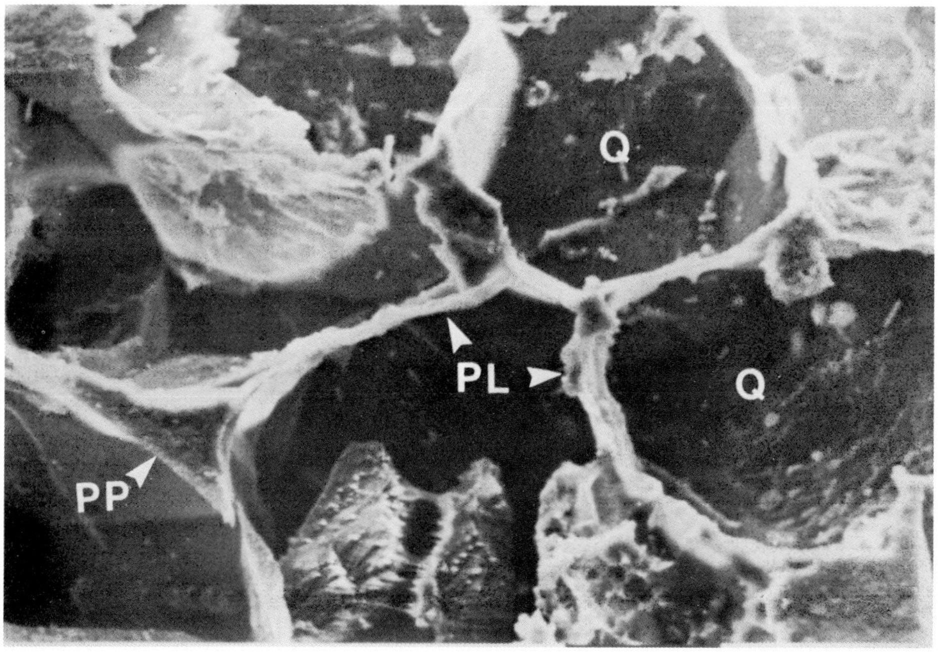

FIG. 14 - Two forms of reduced intergranular porosity

(1) reduced intergranular porosity of lamellar shape that is part primary and part secondary in origin; (2) reduced intergranular pores of polygonal shape are of secondary origin resulting from the dissolution of carbonate cement after the intergranular spaces had been reduced by diagenesis. Scanning-electron photomicrograph of epoxy replica of porosity. Lower Cretaceous, Mackenzie Delta, Canada. PL = lamellar pores; PP = polygonal pores; Q = quartz grains removed by treatment with hydrofluoric acid.

followed by dissolution of the siderite). In most instances, secondary enlarged intergranular pores differ in shape from regular or reduced primary intergranular pores. Pore boundaries are predominantly concave and the adjacent grains are often corroded.

Pores of enlarged intergranular texture constitute a major portion of secondary sandstone porosity. They may occur either in the partial or in the complete mode. A single textural-genetic process may open both the precursor intergranular space and its enlargement (e.g. dissolution of sedimentary intergranular matrix associated with marginal dissolution of adjacent grains). Most commonly, however, enlarged intergranular pores are of hybrid origin (e.g. dissolution of intergranular cement associated with dissolution of replacive minerals at the margins of adjacent grains). Genetic porosity classes and hybrid pore types are the same as those previously indicated for regular intergranular pore texture.

Oversized pore textures. - Pores that exceed the diameter of adjacent grains by a factor of at least 1.2 are, with the exception of fractures, included in the textural group of oversized pores. Two types of secondary oversized pore textures can be differentiated: (1) oversized fabric-selective; and, (2) oversized cross- cutting.

Oversized fabric-selective pore texture describes pores whose shape generally follows fabric elements of the host rock (Figures 16 & 17). The same texture is described as "fenestrate" by Choquette & Pray (1970). It developed selectively at the expense of certain textural elements of the sandstone. Similar pore textures of primary origin are rare but do occur (e.g. keystone vugs formed in sands by escaping air).

FIGS. 15 to 17 - Examples of secondary intergranular and oversized pore textures

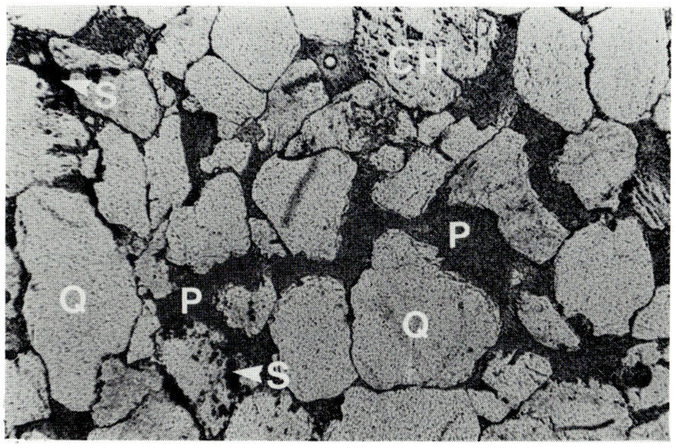

FIG. 15 - secondary pores of enlarged intergranular texture resulting from dissolution of siderite cement and replacive siderite. In places the enlarged intergranular pores are connected and form channels. Half crossed nicols. Jurassic, Beryl Field, North Sea. P = porosity, Q = quartz grains, CH = chert grains, S = siderite cement.

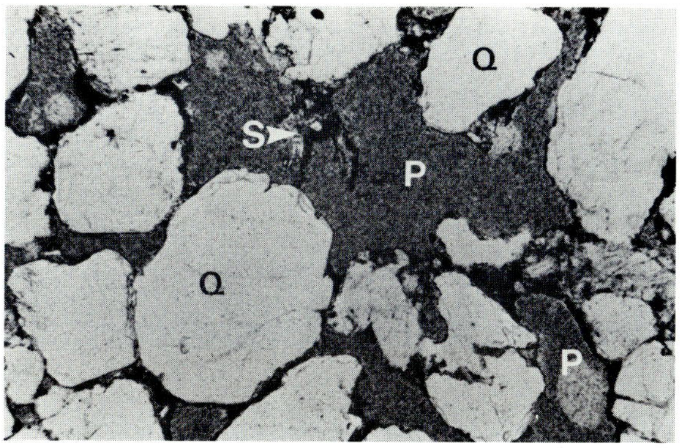

FIG. 16 - Secondary pore of oversized fabric-selective texture resulting from the dissolution of replacive siderite and intergranular siderite cement. Polarizer only. Jurassic, Beryl Field, North Sea. P = porosity; Q = quartz grains; S = siderite remnant (high relief).

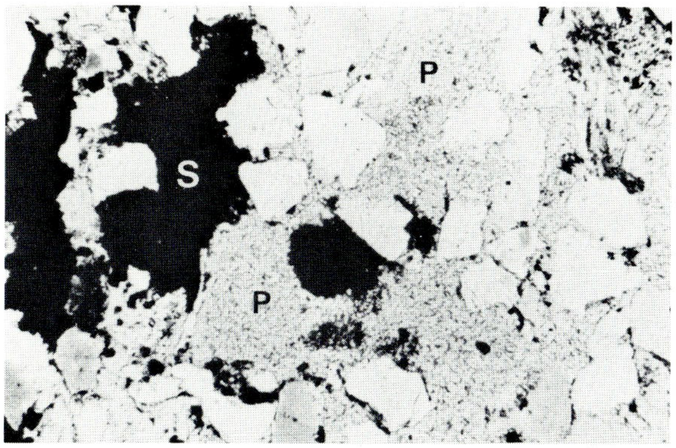

FIG. 17 - Secondary pores of oversized fabric-selective texture resulting from dissolution of sedimentary siderite grains and siderite matrix. Polarizer only. Jurassic, Statfjord Field, North Sea. P = porosity; S = microcrystalline siderite.

Secondary pores of oversized fabric-selective texture are fairly important and fall into four genetic classes: (1) shrinkage (e.g. shrinkage of oversized patches of highly shrinkable matrix); (2) dissolution of sedimentary material (e.g. dissolution of intergranular matrix and adjacent grains); (3) dissolution of cement (e.g. dissolution of cement filling oversized pores); and, (4) dissolution of replacement (e.g. dissolution of carbonate that replaced matrix and adjacent grains). Hybrid pores are common and may represent any combination of these classes (e.g. dissolution of intergranular cement and adjacent grains). Oversized fabric-selective pore texture is also frequently of hybrid primary/secondary origin (e.g. primary intergranular pores combine with secondary porosity created by dissolution of soluble grains).

Oversized crosscutting pore texture is characterized by pore boundaries that cut across the fabric elements of the host rock. Such a texture cannot be of primary origin.

Pores of oversized crosscutting texture are extremely rare but can be valuable as a criterion to establish diagenetic sequences. They result mainly from the dissolution of minerals that indiscriminately replaced patches of the sandstone and, subordinately, from the dissolution of cement following the cementation of first cycle crosscutting pores.

Moldic pore textures. - Pores that show characteristic outlines of the texture of their precursors comprise, with the exception of fractures and intergranular pores, the textural group of moldic pores. Three types of moldic pore textures can be found: (1) grain molds; (2) cement molds; and, (3) replacement molds. Molds formed by dissolution can be either partial or complete. Shrinkage can only create partial molds. Moldic pore textures are always of secondary origin.

Grain mold pore texture forms where secondary porosity is created selectively at the expense of grains which are encased by other grains, matrix, and/or cement or, by minerals replacing the above (Figures 22 & 25).

Pores of grain mold texture are very common and form a considerable percentage of secondary sandstone porosity. Four genetic classes can be recognized: (1) shrinkage (e.g. partial molds created by shrinkage of glauconite grains); (2) dissolution of sedimentary grains; (3) dissolution of cement that filled a first cycle grain mold thus creating a second cycle grain mold; (4) dissolution of replacive minerals (e.g. dissolution of anhydrite that selectively replaced carbonate grains). Molds of mixed primary/secondary origin may occur (e.g. dissolution of a grain with primary intragranular porosity).

Cement mold pore texture is the result of at least three successive diagenetic changes: (1) the precursor cement filled part of a pore developing euhedral crystal faces or an otherwise characteristic morphology; (2) the remaining pore was later filled by another material encasing the precursor cement; and (3) the porosity creating process selectively affected the precursor cement.

Pores of cement mold texture are rare but provide a good criterion for establishing diagenetic sequences. Three genetic classes can be differentiated: (1) shrinkage creating partial molds; (2) dissolution of cement; and, (3) dissolution of minerals that selectively replaced the characteristic cement textures.

Replacement mold pore textures originate if replacive minerals develop euhedral crystal faces or an otherwise characteristic morphology and if, subsequently, a porosity creating process selectively affects the replacement.

Pores of replacement mold texture are rare but are useful for establishing diagenetic sequences. Three genetic classes can be observed: (1) shrinkage creating partial molds; (2) dissolution of cement filling a first cycle replacement mold; and, (3) dissolution of replacive minerals (e.g. dissolution of euhedral anhydrite crystals replacing carbonate matrix and grains).

Intra-constituent pore textures. - The intra-constituent pore textures include all pores within individual textural constituents. Intra-constituent pores range in size from a fraction of a micron to nearly the diameter of the host constituent. Four types of secondary intra-constituent pore textures can be recognized: (1) intragranular; (2) intra-matrix; (3) intra-cement; and, (4) intra-replacement.

Intragranular pore texture results from processes creating porosity within grains rather than only at the grain margins. Grains with a high percentage of intragranular porosity can be termed "honeycombed grains" (Figures 28 & 29). Such grains have been described by Rowsell and De Swardt (1974) as "skeleton grains" and by Fox et. al. (1975) as "leached grains." They are clearly of secondary origin. In grains containing only a relatively small amount of intragranular porosity the nature of the porosity is more difficult to ascertain because many grain types may have some primary intragranular porosity. Both submicroscopic pore diameters (less than 4 microns) and microscopic pore diameters are common in this type of secondary pore texture.

Secondary pores of intragranular texture are fairly

abundant. They result from the following processes: (1) internal shrinkage of grains (e.g. shrinkage of a glauconite grain); (2) internal dissolution of grains (e.g. dissolution of siderite in a sideritic chert grain); (3) internal selective dissolution of cement that may have occupied either primary or secondary intragranular voids (e.g. dissolution of carbonate cement filling primary pores in a chert grain); and, (4) internal selective dissolution of minerals that replaced part of a grain (e.g. dissolution of carbonate that incompletely replaced a feldspar grain). Some grains may contain both primary and secondary intragranular porosity.

Intra-matrix pore textures develop within matrix in a similar fashion to that described for intragranular pore textures (Figure 31). Much of the intra-matrix porosity is submicroscopic (less than 4 microns in diameter). In many instances it is difficult or even impossible to differentiate between primary and secondary intra-matrix porosity.

Secondary pores of intra-matrix texture are fairly common. The same genetic classes are encountered as listed for intragranular pores. Coexistence of primary and secondary porosity is common.

Intra-cement pore textures originate through processes affecting the interior of cements (Figure 30). Intra-cement pores larger than 20 microns in diameter are in most cases of secondary origin as primary intra-cement pores of that size are rare. However, smaller intra-cement pores are commonly of primary origin, as for example in authigenic clay cements. Secondary and primary intra-cement pores commonly coexist. They are often difficult to differentiate from each other if they are of similar size.

Secondary pores of intra-cement texture rarely contribute a significant amount of sandstone porosity. However, they do provide a good criterion for establishing the sequence of diagenetic events. Intra-cement pores develop through the following processes: (1) internal shrinkage of cement (e.g. shrinkage of glauconite cement); (2) internal incomplete dissolution of cement; (3) internal selective dissolution of minerals that replaced part of the cement (e.g. selective dissolution of calcite that replaced the interior of chert cement).

Intra-replacement pore texture originates in the interior of replacive rock constituents in a similar style to that described for intra-cement pore texture. However, pores inside replacive constituents can also be inherited from the precursors (e.g. replacement of a grain with intragranular porosity) or they may be created during the replacement (e.g. replacement of a feldspar grain by porous kaolinite). The pore size ranges from fractions of a micron to dimensions close to the diameter of the replacement textures.

Secondary pores of intra-replacement texture usually form only very small volumes of porosity in sandstones. They develop through the following processes: (1) internal shrinkage of replacive mineral (e.g. shrinkage of replacive glauconite); (2) internal selective dissolution of sedimentary material during replacement process (e.g. dissolution of feldspar during replacement by kaolinite); (3) internal selective dissolution of cement within replacive textures; and, (4) internal dissolution of the replacive minerals.

Fracture pore textures. - Parting and separation of rocks or rock constituents along certain planes creates fractures. Dissolution of material that subsequently filled fractures gives rise to "reopened" fractures. Original and reopened fractures are included in the group of fracture pore textures. Three types of fracture pore textures can be distinguished in sandstones: (1) open rock fractures; (2) open grain fractures; and, (3) open intergranular fractures.

Open rock fractures have a plane of parting that extends over more than a single grain or a single intergranular space. Rock fractures often follow grain boundaries, but in well indurated sandstones they may cut through individual sand grains. They may also cut through matrix, cement, replacement, and other diagenetic fabrics. For this reason they are excellent tools for the investigation of diagenetic sequences (e.g. a stylolite cut by an open fracture predates fracturing). Shape, orientation and size of rock fractures may provide clues to their origin. Open rock fractures in sandstones range in size from microscopic hair fractures less than a millimeter long to fissures of considerable length and width.

Open rock fractures are a very common textural type of porosity but their main importance is the permeability they impart to many sandstone formations. Four genetic classes of open rock fractures can be identified: (1) original fractures; (2) reopened fractures resulting from dissolution of sedimentary material (e.g. filling of a fissure during exposure followed by dissolution); (3) reopened fractures resulting from dissolution of cement; and, (4) reopened fractures resulting from dissolution of replacive minerals. Filled fractures may be partially or completely reopened by dissolution.

Open grain fractures are characterized by planes of parting that are confined to individual grains (Figure 32). Fractures cutting completely through grains are clearly of secondary origin whereas fractures traversing only part of a grain may be either primary or secondary in origin.

The incidence of open grain fractures varies considerably depending on the diagenetic history of the sandstones. They contribute only an insignificant amount of porosity but are of diagnostic value for developing an understanding of diagenesis and porosity history. Three genetic classes of open grain fractures occur: (1) original fractures; (2) reopened fractures resulting from dissolution of cement; and, (3) reopened fractures resulting from dissolution of minerals that replaced fracture cement.

Open intergranular fractures are confined to constituents in single intergranular spaces. They occur in intergranular matrix, in intergranular cement, and in intergranular replacive minerals. They are always of secondary origin. The fractures may dissect the entire intergranular space or traverse only part of it.

Open intergranular fractures are rare but provide excellent evidence for establishing sequences of diagenetic events including both mechanical and chemical compaction of the grain framework. The same genetic classes of open intergranular fractures exist as are listed for grain fractures.

Synopsis of porosity texture. - It is evident from the preceeding discussion that any textural type of secondary porosity can belong to at least three genetic porosity classes including, in every case, voids after authigenic replacement and voids after authigenic cement. In addition, each textural type of secondary porosity includes hybrid pores containing two or more genetic classes. Most pore types can also be of mixed primary and secondary origin.

The majority of secondary sandstone porosity is found in textures that are either identical to, or resemble to some degree, the textures of primary sandstone porosity. However, a considerable portion of secondary sandstone porosity is found in textures that are distinctly different. As will be discussed later, these textures provide evidence for the identification of secondary porosity.

Pittman (1979) defines three geometric-textural classes of sandstone porosity: (1) intergranular porosity; (2) dissolution porosity; and, (3) microporosity. Secondary sandstone porosity is not restricted to the above class of dissolution porosity but is also common in the classes of intergranular porosity and microporosity.

It is readily apparent that effective secondary porosity (porosity that gives rise to appreciable permeability) is found in sandstones of any primary texture including unwinnowed sandstones that never possessed any effective primary porosity. Although effective secondary porosity may simply add to existing effective primary porosity it is commonly found in sandstones that have lost their effective primary porosity.

Porosity of a sandstone can only be increased by fracturing, shrinkage or dissolution if the grain or crystal framework of the sandstones is sufficiently strong to prevent immediate collapse. For this reason there is a limit to the amount of secondary porosity that can develop in a given sandstone lithology under specific conditions of stress because the supportive strength decreases with increasing porosity and the framework will ultimately fail.

Secondary sandstone porosity may be reduced and texturally modified by the same processes that reduce primary porosity:

(1) authigenic cementation; (2) dissolution at grain contacts; and, (3) mechanical compaction. These textural processes not only reduce the size of pores and change their shape but frequently obscure the evidence for their secondary origin.

RECOGNITION

It is important to realize that most secondary sandstone porosity closely resembles or exactly mimics primary porosity. This is the reason why secondary sandstone porosity has largely remained unnoticed until recently. It follows that, in the study of secondary sandstone porosity, it is necessary to develop readily observable diagnostic criteria for recognizing or suspecting its presence. Additionally, a suitable analytical approach must be chosen for verifying suspected cases and for recognizing the quantity, texture, origin, and textural modification of secondary porosity.

<u>Petrographic Criteria</u>. - In many instances, the secondary nature of sandstone porosity can be identified by using petrographic criteria that are easily observable in thin sections under the petrographic microscope. Some of these can also be observed, on occasion, in hand specimens. The following eight petrographic criteria appear to be the most useful (Figure 18): (1) partial dissolution; (2) molds; (3) inhomogeneity of packing; (4) oversized pores; (5) elongate pores; (6) corroded grain margins; (7) intra-constituent pores; and, (8) fractured grains.

<u>Partial dissolution</u> of soluble constituents is by far the most conclusive criterion and it is also very common. Dissolution of soluble sedimentary and authigenic constituents is, in many instances, incomplete and patches of remnant material occur adjacent to pores (Figure 21). Frequently the

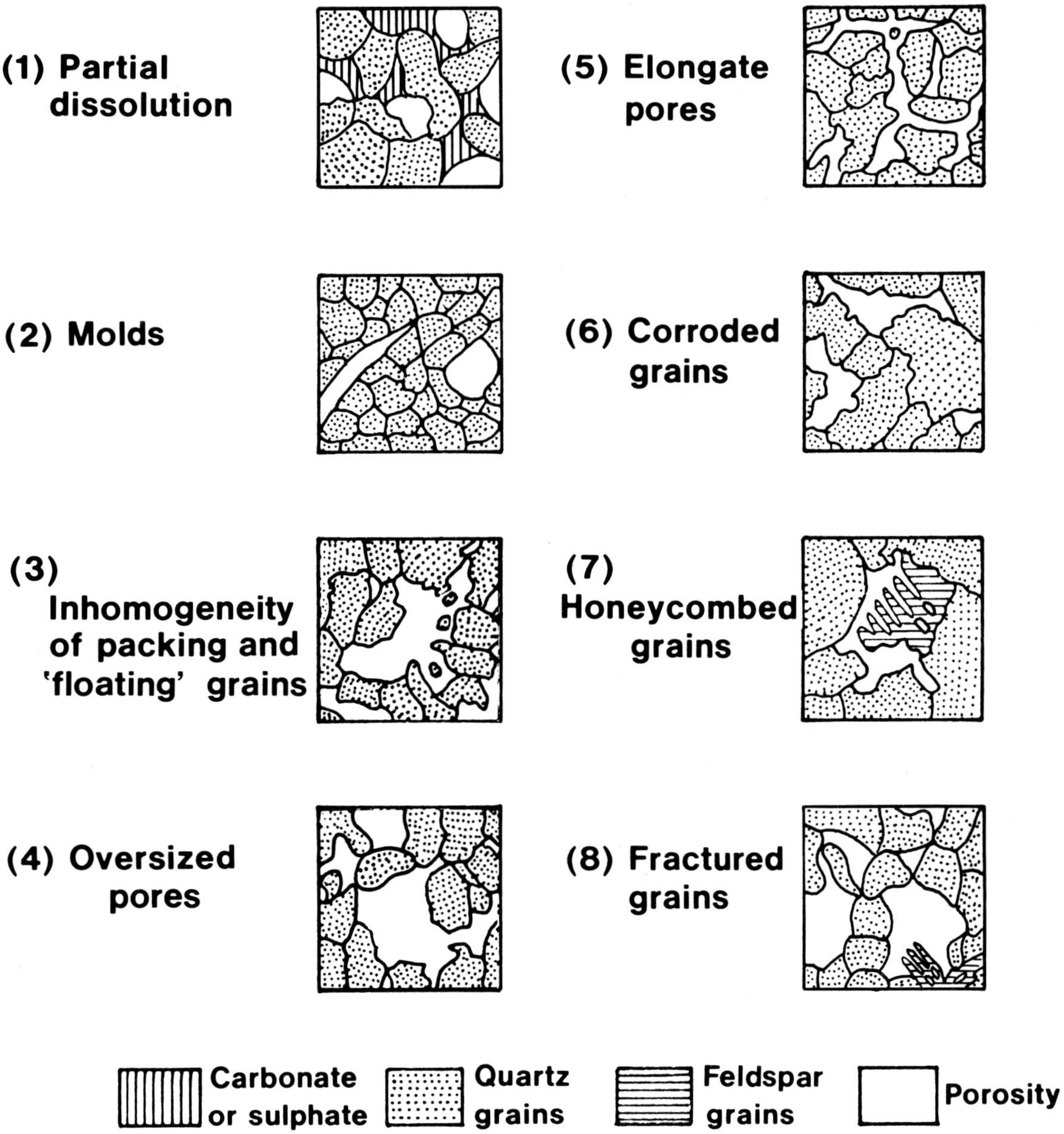

FIG. 18 - Petrographic criteria for recognition of secondary sandstone porosity.

remnants have a corroded appearance. Care must be taken to determine the partial dissolution of cements because incomplete cementation of porosity may produce very similar textures. Indicators for the partial dissolution of authigenic cements are: (1) corroded surfaces; (2) uniform extinction of isolated remnants of poikilotopic cement crystals; (3) intra-cement pores; and, (4) disrupted growth zonations in the cement. In some instances, partial dissolution produces large diagenetic structures that can be observed in hand specimens without the aid of a microscope (Figure 19).

Molds of sedimentary particles, cement and replacive minerals (Figures 22 & 23) are excellent diagnostic features if the outlines of the moldic pores clearly portray the characteristic shapes of their precursors. Grain molds are especially useful because of their frequent and widespread occurrence. Commonly, molds can be identified directly in drill cuttings and hand specimens using only a wellsite binocular microscope or a handlens. Molds of large debris (Figure 20) and large authigenic textures are visible to the naked eye.

Inhomogeneity of packing is a very useful and common criterion. Sandstones displaying marked inhomogeneity with areas of loosely-packed grains and high intergranular porosity adjacent to areas of tightly-packed grains and little or no porosity (Figure 24) reflect an original inhomegeneity in the distribution of soluble matrix or cement. The tight packing is the result of mechanical compaction and chemical grain interpenetration in areas lacking intergranular matrix or cement. In adjacent areas the presence of intergranular constituents prevented compaction and grain interpenetration. Later dissolution of soluble cement or matrix created the secondary porosity between the loosely-packed grains. Individual grains may straddle the boundary between open and

FIGS. 19 & 20 - Macroscopic criteria for secondary sandstone porosity

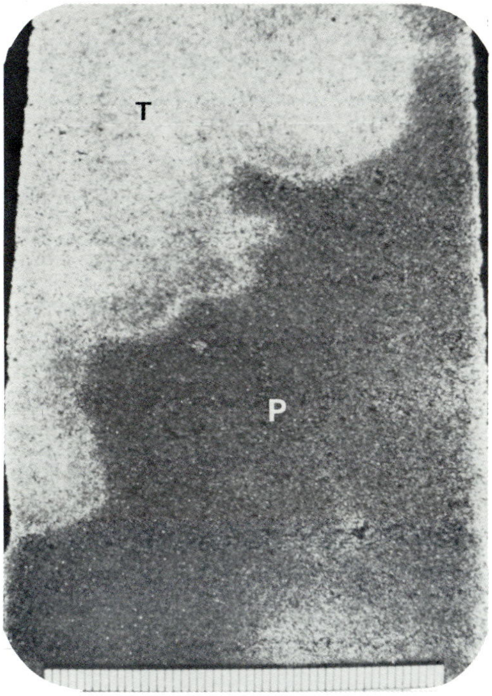

FIG. 19 - Example of dissolution front: contact between sandstone with high secondary porosity resulting from dissolution of calcite cement (dark) and non-porous sandstone containing a high amount of calcite cement that filled primary porosity (light). Core photograph. Lower Cretaceous, Mackenzie Delta, Canada. P = sandstone porosity app. 25%; T = tight sandstone with 25% calcite cement.

FIG. 20 - Example of macroscopic vuggy porosity: conglomeratic sandstone contains high vuggy porosity resulting largely from the dissolution of siderite clasts. Core photograph. Lower Cretaceous, Mackenzie Delta.

FIGS. 21 to 23 - Examples of criteria for recognition of secondary sandstone porosity

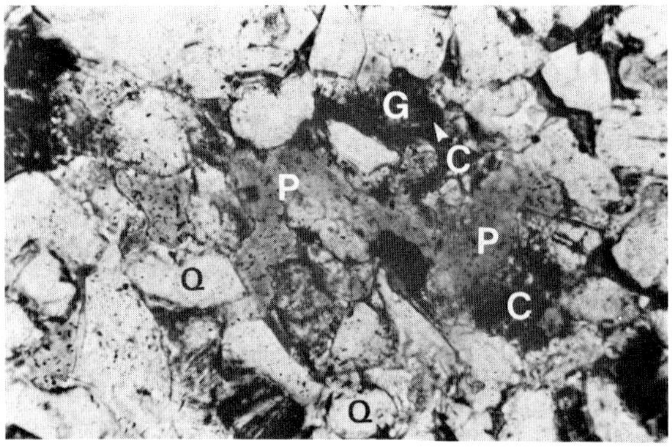

FIG. 21 - Example of partial dissolution: high secondary porosity created by partial dissolution of calcite cement and calcite replacing glauconite. Secondary porosity consists of intergranular pores and oversized fabric selective vugs. Polarizer only. Cretaceous, Scotian Shelf, Canada. P = porosity; C = calcite; Q = quartz grains; G = glauconite.

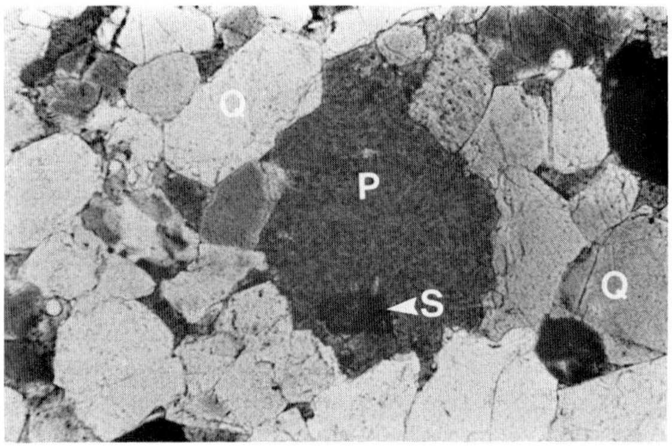

FIG. 22 - Example of grain mold: mold of large siderite grain resulting from nearly complete dissolution of siderite. Mold surrounded by welded quartz grains. Half crossed nicols. Lower Cretaceous, Mackenzie Delta, Canada. P = porosity; S = siderite remnant (high relief); Q = quartz grains.

FIG. 23 - Example of grain mold: grain mold resulting from dissolution. The original grain was coated by clay cement. The remaining rim of clay cement has become the shell for the grain mold. Half crossed nicols. Lower Cretaceous, Mackenzie Delta, Canada. P = porosity; Q = quartz grains; CH = chert grain with intergranular porosity; M = grain mold; S = siderite remnant.

FIGS. 24 to 26 - Examples of criteria for recognition of secondary sandstone porosity

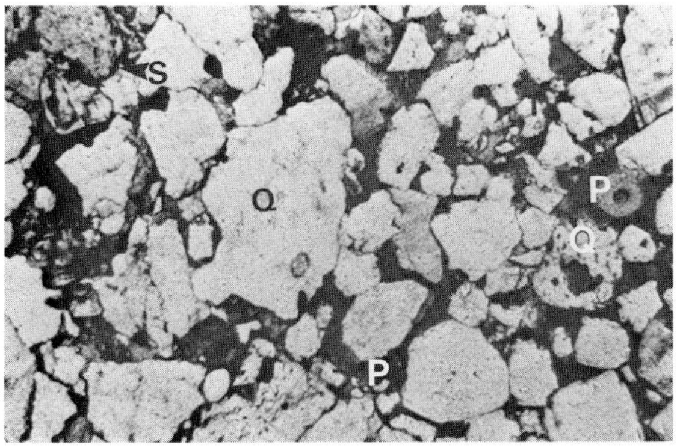

FIG. 24 - Example of inhomogeneity of packing: (1) Loose packing with high effective porosity (approx. 25 vol. %). Porosity was created by dissolution of siderite. (2) Tight packing with no effective porosity. Primary porosity lost through chemical compaction. Polarizer only. Jurassic, Beryl Field, North Sea. P = porosity; Q = quartz and quartzite grains; S = stylolite.

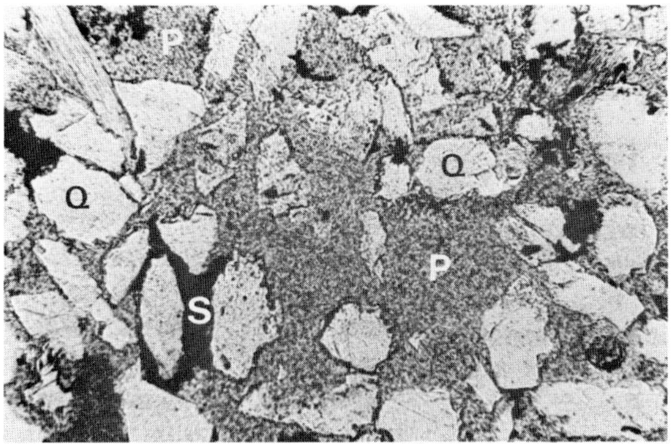

FIG. 25 - Example of oversized pores and floating grains; exceedingly high secondary porosity (approx. 40 volume %) resulted from the dissolution of siderite matrix, siderite grains and siderite that replaced the margins of sand grains. Polarizer only. Jurassic, Statfjord Field, North Sea. P = porosity; Q = quartz grains; S = siderite.

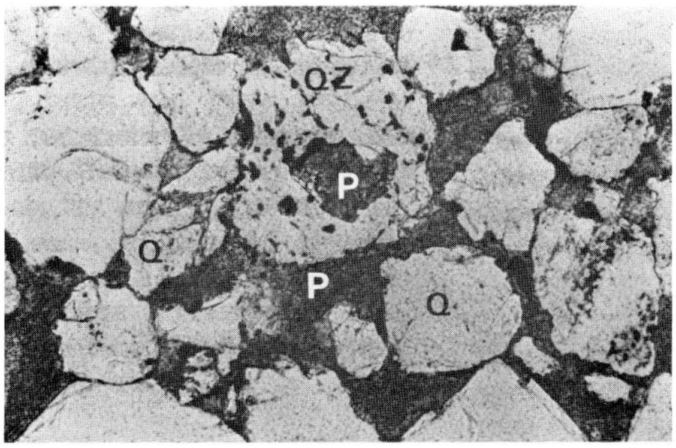

FIG. 26 - Example of corroded grains: detail of Fig. 24. Dissolution of replacive siderite gave rise to corroded margins of grains and intragranular pore in quartzite grain. Polarizer only. P = porosity; Q = quartz grains, in part, fractured; QZ = quartzite grain.

tight packing thereby forming very diagnostic fabrics. For example, one part of a mica fragment may be strongly deformed between compacting grains, while the other part of the mica retained its original shape and forms the boundary of a pore in an area of open packing. The latter pore resulted from dissolution of eogenetic intergranular cement which protected part of the mica from deformation during compaction.

Inhomogeneity of packing can be best seen in thin sections oriented normal to bedding. However, it can often be recognized in drill cuttings, cores or outcrop samples by means of a stereo microscope, handlens and, on occasion, the naked eye.

Oversized pores that have a significantly larger diameter than that of adjacent grains are common and provide good evidence for secondary porosity (Figure 25). Most oversized pores are fabric selective and result mainly from the selective dissolution of soluble sedimentary clasts and lenses of matrix or of their replacive successors.

Oversized pores frequently occur together with inhomogeneity of packing and may also form channels, in which cases they clearly indicate the presence of secondary porosity. However, as mentioned earlier, oversized pores can also be created by sedimentary processes. Oversized pores can often be recognized without the aid of thin sections.

Elongate pores, if numerous, indicate the presence of secondary porosity. Replacement processes often operate at the margins of tightly-packed grains. If these replacive minerals are dissolved, the sandstone will exhibit a high incidence of elongate intergranular pores (Figure 27). This criterion has to be used with caution because primary elongate pores are

FIGS. 27 to 29 - Examples of criteria for recognition of secondary sandstone porosity

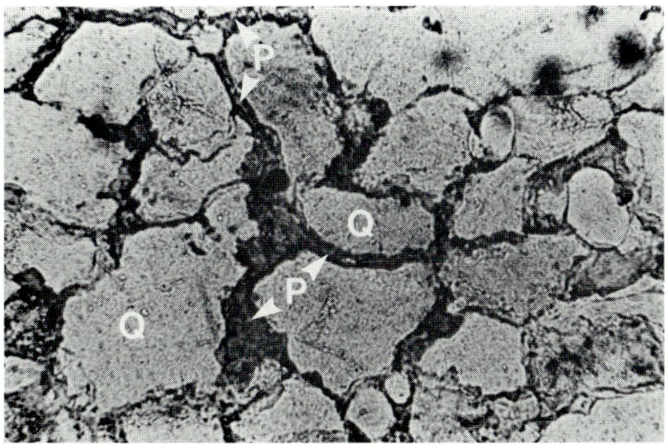

FIG. 27 - Example of elongate pores: the elongate pores resulted from marginal replacement of quartz grains by carbonate minerals along irreducible lamellar pores followed by dissolution of carbonate. Polarizer only. Cambrian, Northwest Territories, Canada. P = porosity; Q = quartz grain.

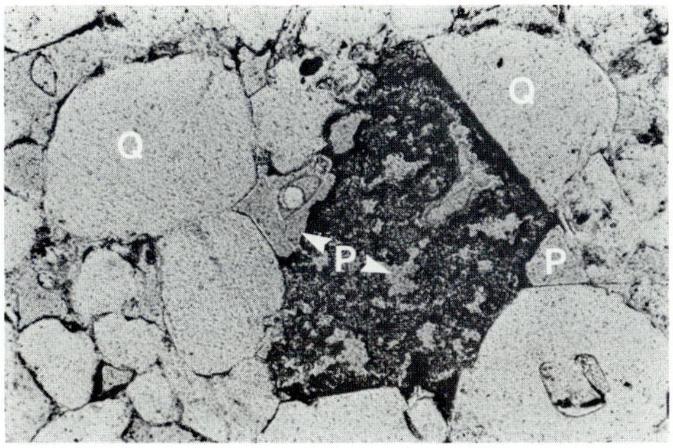

FIG. 28 - Example of honeycombed grain: dissolution of primary carbonate siderite inclusion in chert created high intragranular porosity in chert grain. Polarizer only. Lower Cretaceous, Mackenzie Delta, Canada. P = secondary intragranular and intergranular porosity; Q = quartz grains.

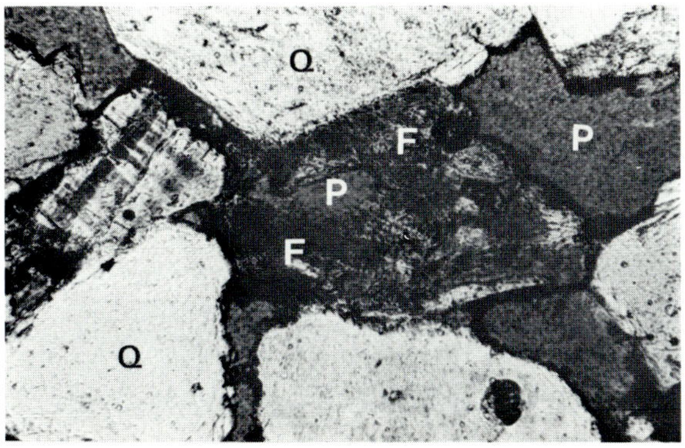

FIG. 29 - Example of honeycombed grain: intragranular porosity in feldspar grain resulted from dissolution of replacive carbonate. Half crossed nicols. Cretaceous, Scotian Shelf, Canada. P = intergranular and intragranular porosity; Q = quartz grains; F = feldspar (honeycombed).

common in sandstones with abundant platy grains and in well packed and very fine-grained sandstones with grains of high angularity. Elongate pores are best observed in thin section.

Corroded grains adjacent to pores are very frequently a useful criterion for secondary porosity. They generally are associated with enlarged intergranular pores (Figure 26) and result mainly from dissolution of soluble minerals that unevenly replaced the margins of sand grains. Quartz grains with corroded syntaxial overgrowths of authigenic quartz are reliable indicators of secondary porosity. However, mildly corroded grains may be of sedimentary origin. Additionally, unrounded, fine sand grains may easily be mistaken for corroded grains. The criterion of corroded grain margins, therefore, must be applied with care. The reliability generally increases with increasing severity of the corrosion, because severely corroded grain margins are not likely to survive sedimentary transport.

The observation of diagenetic corrosion of grains is best accomplished in thin section. However, corroded grains can be recognized in drill samples of medium and coarse-grained sandstones with the well site binocular microscope. In the absence of cores, loose corroded grains may be the only direct evidence of the nature of porosity, if sandstones with high secondary porosity are encountered in wells.

Intra-constituent pores are fair to excellent indicators of secondary porosity. Among constituents with internal pores, honeycombed grains provide conclusive evidence for post-depositional leaching, because they would quickly disintegrate during sedimentary transport. They are mainly the result of leaching of carbonates which partially replaced sand grains - typically feldspars, clay pellets and chert fragments

(Figures 28 & 29). Honeycombed grains are easy to recognize in thin section and have been mentioned as evidence for secondary porosity by several authors, including Rowsell and De Swardt (1974), Stanton and McBride (1976), Hayes (1979), and Pittman (1979). Other types of intra-constituent pores that occasionally provide evidence for secondary porosity include: (1) isolated intragranular pores; (2) intra-cement pores (Figure 30); (3) intra-matrix pores (Figure 31); and, (4) intra-replacement pores. These intra-constituent pores generally must be examined in thin section in order to establish their secondary nature.

Open grain fractures are a very useful criterion in sandstones that experienced grain welding and syntaxial overgrowth of grains. They generally represent framework failure caused by the generation of a high percentage of secondary porosity during burial (Figure 32). Clearly open grain fractures formed after grain welding and syntaxial overgrowth had taken place since pre-existing grain fractures would have been healed during the development of syntaxial overgrowths. Grain fractures are best examined in thin section.

Method of analysis: For a thorough analysis of secondary sandstone porosity the following sequence of steps should be taken: (1) recognize the presence of secondary porosity; (2) identify the textural types of secondary porosity present; (3) determine the textural-genetic origin of the secondary porosity and any subsequent textural modification; (4) ascertain the abundance of secondary porosity and the abundance and texture of any coexisting primary porosity; and, (5) determine the sequential order of diagenetic changes that affected the sandstone, including the alteration of primary porosity and the generation and alteration of secondary

FIGS. 30 to 32 - Examples of criteria for recognition of secondary sandstone porosity

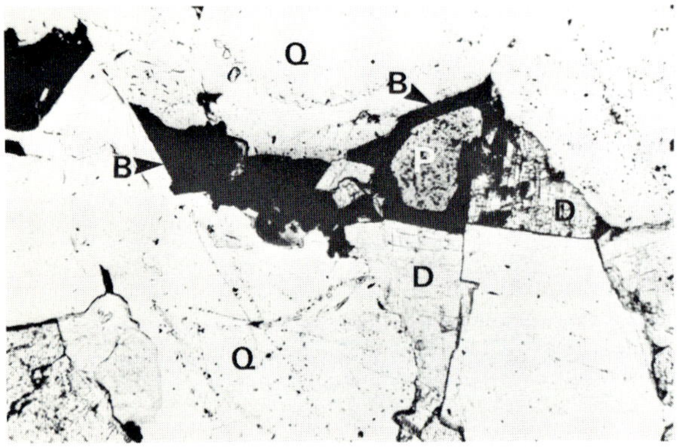

FIG. 30 - Example of intra-cement pores: partial dissolution of dolomite cement created intra-cement pores which were later filled by bitumen. Cambrian, Northwest Territories, Canada. P = porosity; D = dolomite cement and replacive dolomite; B = bitumen; Q = quartz grains.

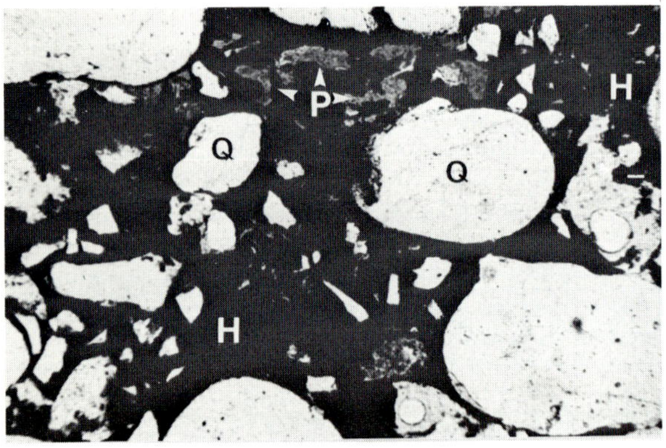

FIG. 31 - Example of intra-matrix pores: shrinkage og glauconite and/or hematite created intra-matrix pores. Polarizer only. Cambrian, Northwest Territories, Canada. P = porosity; H = hematite matrix which probably replaced glauconite matrix; Q = quartz grains.

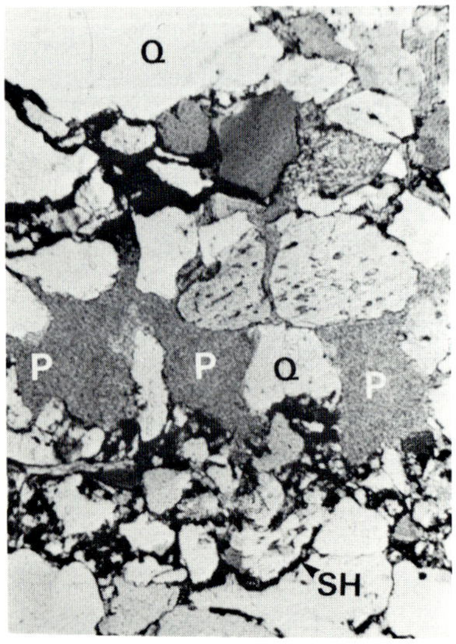

FIG. 32 - Example of open grain fracture: dissolution of carbonate created secondary porosity. Grain fracturing resulted from increased stress in a pillar of grains supporting the roof of large adjacent secondary pores. Jurassic, Beryl Field, North Sea. Half crossed nicols. P = secondary porosity; Q = quartz; SH = shale lamina.

porosity.

The stratigraphic distribution of porosity may, in some instances, suggest secondary origin. For example, a plot of porosity versus depth may show a sudden increase in porosity at great depth beneath a long interval of gradually declining porosities (Proshlyakov, 1960). Certain physical reservoir characteristics may also provide indirect evidence for secondary porosity (Schmidt et. al., 1977).

Direct evidence for secondary porosity can, on occasion, be observed macroscopically, or else with the aid of a hand lens or a stereo microscope. In most instances, however, direct evidence can only be obtained by more advanced methods of observation. The occurrence of secondary porosity can usually be detected by examining a few dozen thin sections with a good petrographic microscope and applying the criteria mentioned earlier. The pore system of the sandstones should be thoroughly impregnated with colored epoxy resin before thin sectioning in order to emphasize pores and to prevent damage to pore boundaries during grinding. A combination of as many criteria as can be observed should be taken into account because it may not be possible to arrive at a valid conclusion by using only one or two criteria. Thin section analysis of drill cuttings can produce excellent results since a single thin section may show up to fifty cuttings portraying the entire lithologic spectrum of a sandstone interval.

Conventional petrographic microscopy may not suffice to positively identify the secondary porosity of some sandstones. In such cases cathode luminescence microscopy, exography, scanning-electron microscopy and microprobe analysis will probably permit detection of the secondary porosity.

The accurate identification of the various textural types of secondary porosity in a sandstone unit requires nothing less than a thorough petrographic examination of thin sections representing the entire spectrum of lithologies present. In addition, a study of pore casts is always advisable.

If a significant amount of pores and other textural features measure less than 10 microns then it is necessary to employ scanning-electron microscopy of rock samples and pore casts.

The detection of the textural-genetic origin of secondary porosity requires the petrographic comparison of samples showing fully developed secondary porosity with those showing only incipient development of secondary porosity and those without any secondary porosity. In addition to petrographic microscopy and scanning-electron microscopy of samples and pore casts it is often useful to apply cathode luminescence microscopy and microprobe analysis. In the case of dissolution porosity, comparison of casts of artifical pores with casts of natural pores provides a powerful analytical tool. The casts of artificial pores are obtained by acid treatment of samples containing soluble constituents but devoid of secondary porosity.

Secondary porosity can usually be recognized in sandstones lacking significant amounts of primary porosity. However, in sandstones with significant amounts of both primary and secondary porosity it is often difficult to assess the relative abundance of each form of porosity. This is especially true if primary and secondary porosity coexist in a large percentage of the pores. In such cases it is advisable to study, for comparison, the texture of primary porosity in sandstones of the same original lithology and the same state of compaction but without the development of secondary porosity. In some instances only part of the porosity can be positively recognized as primary and secondary while the origin of the remaining part of the porosity remains undetermined.

The final step in the petrographic analysis of secondary porosity is to establish the sequence of diagenetic events. The generation of secondary porosity may have occurred only one time or several times during the diagenetic history. The textural relationship between secondary porosity and sandstone host provides evidence of the diagenetic changes that affected the rock prior to and following the formation of the secondary porosity. For example, secondary pores which cut across the welded grain fabrics indicate that welding occurred before leaching. On the other hand, secondary pores that are lined by bitumen originated prior to migration. Thus diagenetic sequences can be established and fitted to the geological history of the sandstones.

Much of the evidence for the sequence of diagenetic events in a sandstone unit is established by the careful analysis of the types of textures of the secondary porosity and its genetic origin. For example, the fabric relationships between fractures and diagenetic textures and structures often provides an excellent means of establishing sequences of diagenetic events in sandstone. More evidence may be obtained by comparison with a sandstone unit of similar original composition that experienced less diagenetic alteration.

In addition to the analytical techniques discussed above, the following analytical methods may, on occasion, be useful in retracing the diagenetic history: (1) X-ray diffraction analysis; (2) stable isotope analysis; (3) analysis of fluid inclusion; and, (4) analysis of the thermal alteration of organic matter. In working with all these data it is necessary to strictly apply the principle of multiple working hypotheses.

The petrologic record of many sandstone units does not suffice to conclusively establish their diagenetic history. In these cases, the integration of the petrological data with other geological information such as burial history, compaction history, and the history of fluid migration often results in the correct interpretation of the diagenesis.

GEOLOGICAL OCCURRENCE AND ORIGIN OF SECONDARY SANDSTONE POROSITY

Regimes of diagenesis. - Secondary sandstone porosity is formed and preserved, changed and destroyed in almost any conceivable diagenetic environment. The system of diagenetic regimes that Choquette and Pray (1970) formulated for the study of porosity in carbonate rocks is equally appropriate for investigation of secondary porosity in sandstones and has been adopted in this paper. Eodiagenesis is defined as the regime at or near the surface of sedimentation where the chemistry of the interstitial water is mainly controlled by the surface environment prior to effective burial. (Effective burial is here considered as burial under strata that seal the sandstone from a predominant influence of surface agents on the chemistry of the interstitial water). Mesodiagenesis is the subsurface regime during effective burial. Telodiagenesis represents the regime at or near the surface after effective burial. The amounts of secondary porosity originating in each diagenetic regime differ greatly (Figure 33).

Eodiagenesis produced a minor portion of secondary porosity, mainly as a result of dissolution of sedimentary and eogenetic constituents, although fracturing and shrinkage may be subordinate contributors. In some instances eogenetic

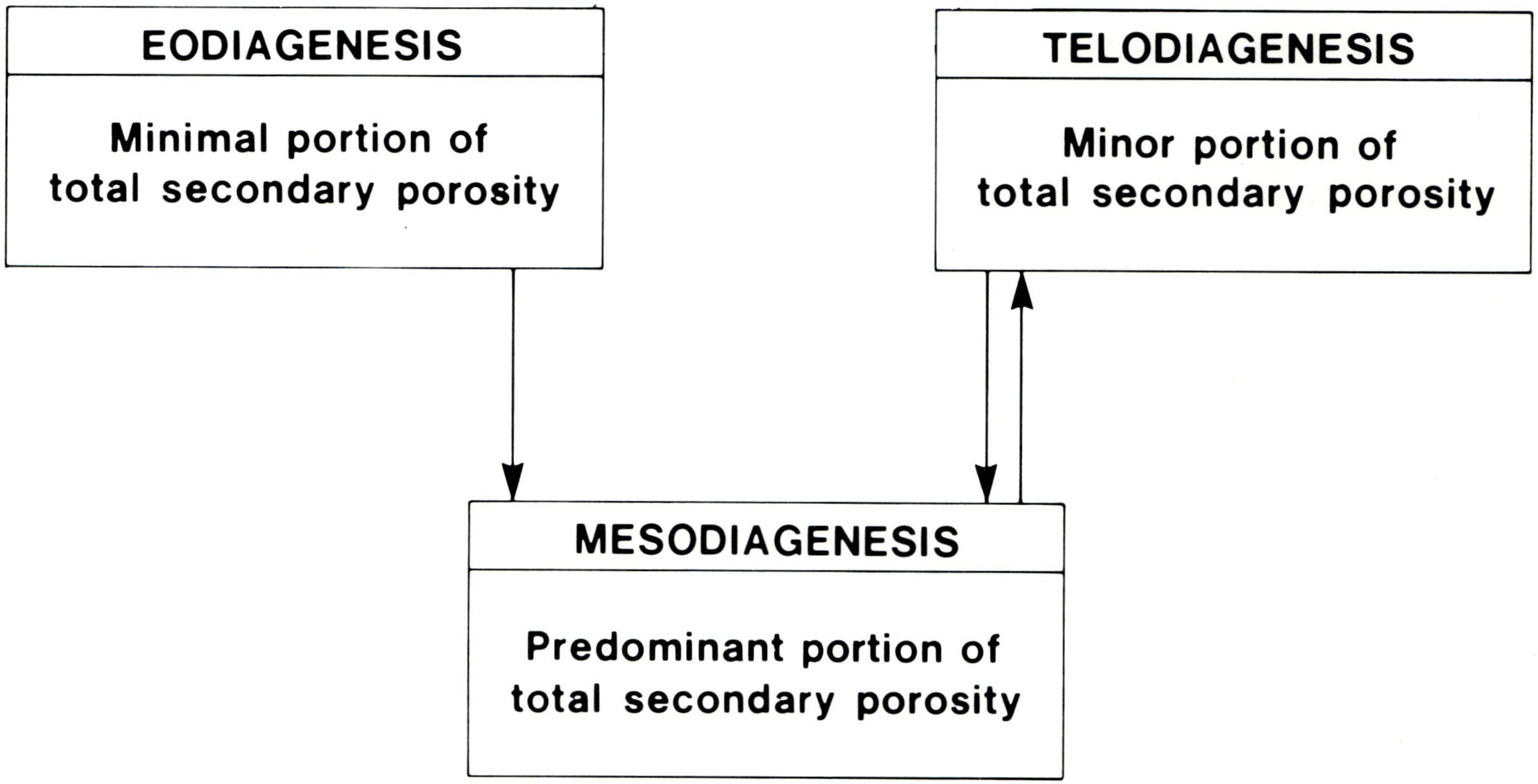

FIG. 33 - Diagenetic regimes of development of secondary sandstone porosity.

processes created considerable secondary porosity, for example, in calcareous quartz sandstones of Pleistocene and Holocene age from the Trucial Coast (unpublished petrographic study by Schmidt).

Mesodiagenesis created by far the largest amount of secondary porosity, and most of it can be attributed to leaching of sedimentary, eogenetic, and mesogenetic carbonate constituents (Hayes, 1979; Schmidt, 1976). Fractures form an important, although subordinate, portion of mesogenetic sandstone porosity. Mesogenetic shrinkage porosity plays a minor role and much of the shrinkage occurs during shallow burial.

Telodiagenesis contributed only a minor share of the total sandstone porosity, mainly as the result of leaching and, subordinately, due to fracturing. This we conclude from the comparison of outcrop samples and subsurface cores of numerous sandstone formations, and from the stratigraphic consideration that only a small percentage of sandstones have undergone a period of telodiagenesis.

Diagenetic processes. - The individual diagenetic processes that result in secondary porosity vary significantly in their quantitative importance (Figure 34).

Fracturing of rocks, grains and intergranular constituents generally creates only insignificant to minor pore volumes. However, rock fractures can, on occasion, form the major porosity type in some sandstone with low matrix porosity.

Shrinkage, if present, usually supplies little porosity although exceptions do occur.

FIG. 34 - Diagenetic processes creating secondary sandstone porosity.

DIAGENETIC PROCESSES		PORTION OF SECONDARY POROSITY
	ROCK FRACTURING	* MINOR
	GRAIN FRACTURING	MINOR
	SHRINKAGE	* MINOR
DISSOLUTION OF	CALCITE	MAJOR
	DOLOMITE	MAJOR
	SIDERITE	MAJOR
	SULFATE	* MINOR
	OTHER EVAPORITES	* MINOR
	SILICATE	VERY MINOR
	OTHER NONSILICATE	VERY MINOR

* MAY BE OF MAJOR IMPORTANCE IN INDIVIDUAL STRATIGRAPHIC UNITS

<u>Dissolution of carbonate minerals</u> created the vast majority of the total volume of secondary sandstone porosity. Calcite, dolomite, and siderite each are the precursors of secondary porosity on a major scale (Figures 35, 36, 37). The dissolved carbonate constituents mainly consist of cements, but also include substantial percentages of sedimentary and replacive carbonate.

<u>Dissolution of sulfate minerals</u> such as anhydrite, gypsum, barite, and celestite is common (Figure 38). This may occur during any stage of diagenesis. The secondary porosity resulting from the removal of sulfates is minor in most sandstone formations. However, it can be of major importance in some instances, for example, in Cambrian sandstones of the Northwest Territories (McDonald, in progress).

<u>Dissolution of nonsulfate evaporite minerals</u> such as halite and sylvite appears to create only a small fraction of secondary sandstone porosity because of the relatively infrequent occurrence of these evaporite minerals in sandstones. Porosity of this origin, however, is easily misidentified and may be much more common than is presently realized. Hartman (1968) recognized a case of leaching of halite that created the reservoir for the Pelahatchie Field in the Gulf Coast region of the United States. Therefore, dissolution of nonsulfate evaporite minerals can, in some instances, develop secondary porosity of major importance. It may take place either in the subsurface or near the surface in contact with meteoric waters.

<u>Direct dissolution of silicate minerals</u> such as quartz or feldspar generally creates little or no secondary porosity. Dissolution of silicate minerals as a rule occurs along grain contacts without creating porosity. However, Hayes (1979)

FIGS. 35 to 37 - Mesogenetic dissolution porosity

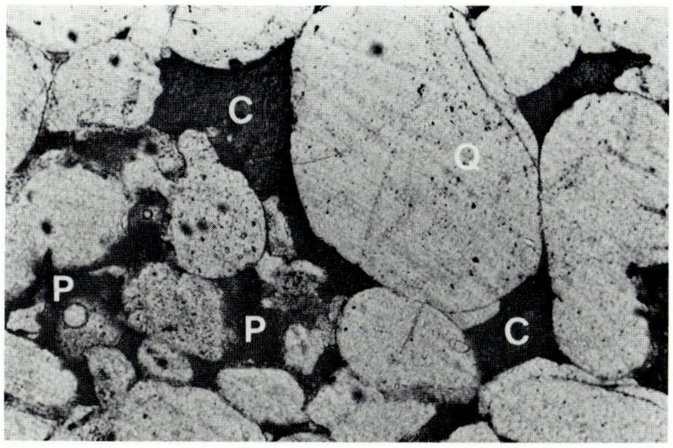

FIG. 35 - Mesogenetic dissolution of poikilotopic calcite cement created secondary porosity in Cambrian quartz arenite, Northwest Territories, Canada. Polarizer only. P = secondary porosity; Q = quartz grains and mesogenetic syntaxial quartz overgrowth; C = mesogenetic calcite cement.

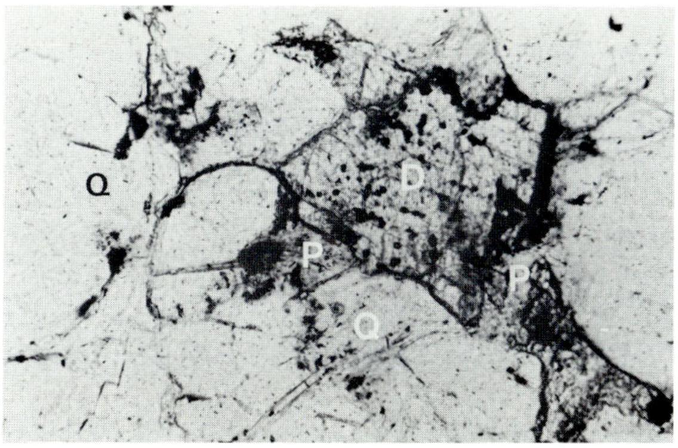

FIG. 36 - Mesogenetic dissolution of dolomite cement created secondary porosity in Lower Cretaceous quartz arenite, Scotian Shelf, Canada. Polarizer only. P = secondary porosity; Q = quartz grains and mesogenetic syntaxial overgrowth; D = dolomite (high relief).

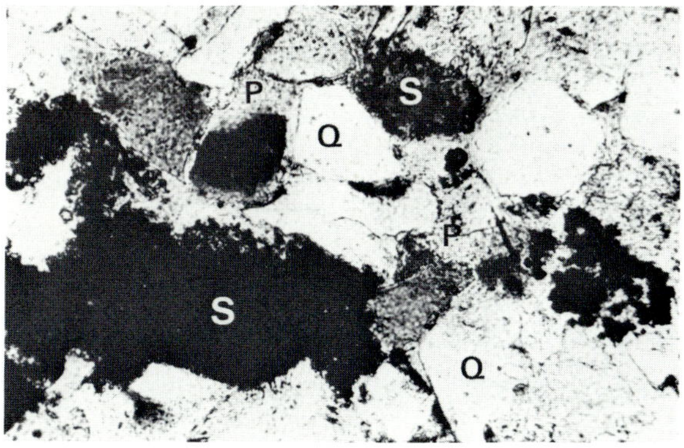

FIG. 37 - Mesogenetic dissolution of siderite created secondary porosity in Jurassic quartz arenite, Statfjord field, North Sea. Half crossed nicols. P = secondary porosity; Q = quartz grains; S = siderite (very high relief).

FIGS. 38 to 40 - Mesogenetic dissolution porosity

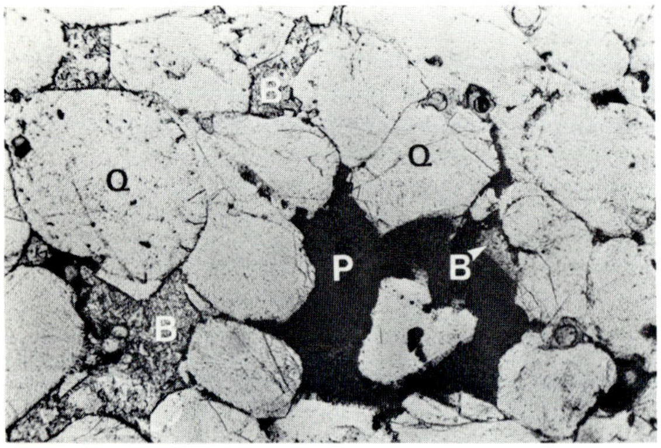

FIG. 38 - Mesogenetic dissolution of poikilotopic barite cement created secondary sandstone porosity in Cambrian quartz arenite, Northwest Territories, Canada. Polarizer only. P = secondary porosity; Q = quartz grains and mesogenetic syntaxial quartz overgrowth; B = barite (moderately high relief).

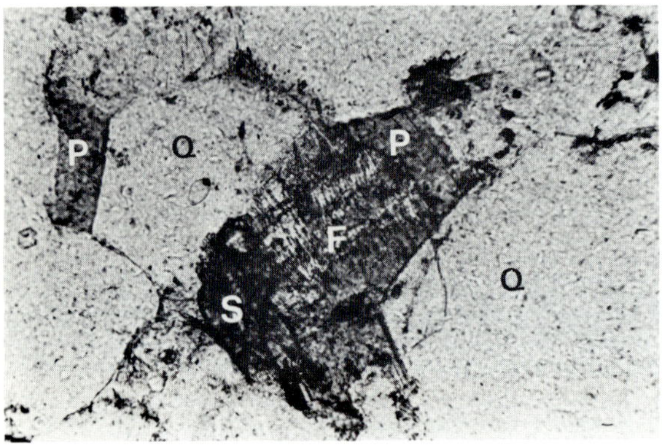

FIG. 39 - Siderite partially replaced feldspar. Subsequent mesogenetic dissolution of replacement siderite created secondary porosity in Lower Cretaceous subfeldspathic arenite, Mackenzie Delta, Canada. Polarizer only. P = secondary porosity; Q = quartz grains and mesogenetic syntaxial quartz overgrowth; F = feldspar (honeycombed grain); S = siderite (very high relief).

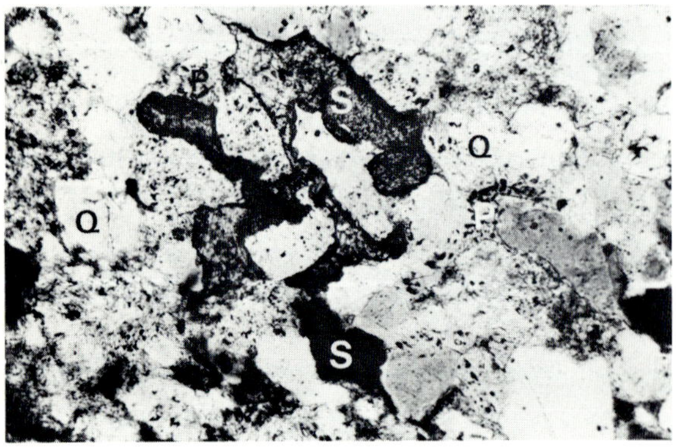

FIG. 40 - Siderite marginally replaced quartz. Subsequent mesogenetic dissolution of replacement siderite and siderite cement created secondary porosity in Triassic quartz arenites, Prudhoe Bay Field, North Slope, Alaska. Half crossed nicols. P = secondary porosity; Q = quartz grains; S = siderite (very high relief).

reports a possible case of significant leaching of silicate minerals in Tertiary sandstones of South Central Alaska. Heald and Larese (1973) attribute the porosity in honeycombed feldspar of Paleozoic sandstones in West Virginia and Ohio to direct dissolution. Their evidence, however, is not conclusive, as leaching of carbonate may be followed by precipitation of carbonate cement. Silicate minerals are often replaced by non-silicate minerals, principally carbonates. These replacive minerals, in turn, are frequently dissolved (Figures 39, 40) and the resulting secondary porosity may be mistakenly attributed to direct leaching of the silicate minerals.

Dissolution of other non-silicate minerals including, for example, phosphate and oxide minerals contributes very little secondary porosity in sandstones.

Causes. - The conditions which bring about the generation of secondary porosity in sandstones are the result of chemical, physicochemical, physical, biochemical, and biophysical processes.

Chemical processes such as dilution of the interstitial water or changes in the ratios of ion species may cause void-creating dissolution at constant temperatures and pressures. The formation of carbonic acid in the interstitial water is a well known example. It causes lowering of the pH and dissolution of carbonate minerals. In the telogenetic regime carbonic acid is derived mainly from atmospheric carbon dioxide. During mesodiagenesis most of the carbon dioxide that forms carbonic acid originates from organic matter undergoing thermomaturation (Figure 41). Tissot et al (1974) demonstrated that this release of mesogenetic carbon dioxide occurs over a wide range of temperature/time exposures. Savkevic (1971)

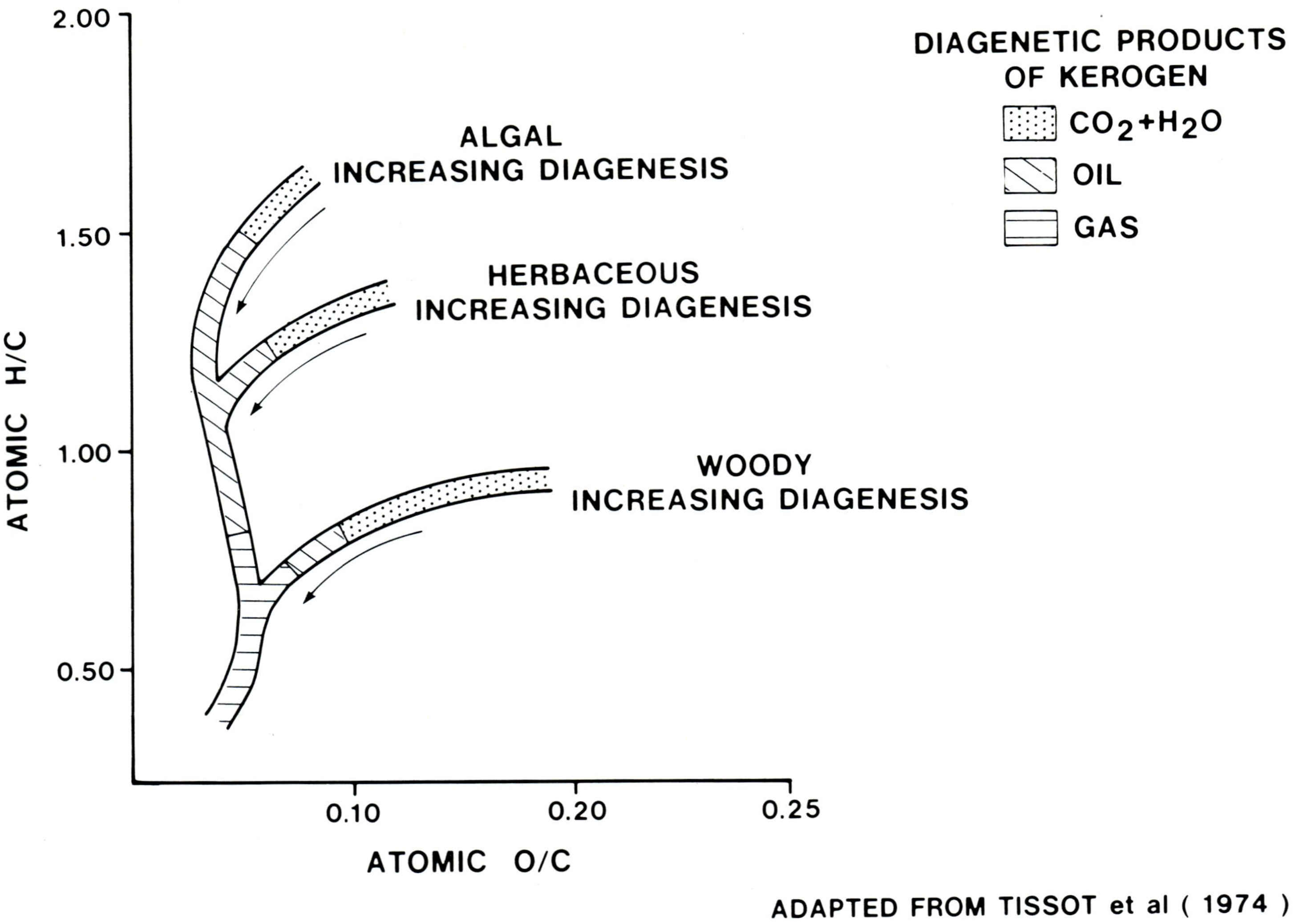

FIG. 41 - Diagenetic evolution of kerogen.

suggested that both the interlayer water which is released from montmorillonite during its mesogenetic transformation to illite and carbon dioxide originating from the oxidation of organic matter could cause subsurface brines to become so undersaturated in respect to carbonate minerals that the latter are dissolved. Parker (1973) concluded that carbon dioxide originating from a plutonic intrusion was responsible for the mesogenetic leaching of carbonate which created the secondary sandstone porosity in the Smackover sandstones of the Thomasville Field, Mississippi. Hydrogen sulfide may also cause mesogenetic dissolution of carbonate (Parker, ibid.).

Physicochemical processes can cause dissolution and shrinkage because changes in temperature and pressure affect the solubility of minerals if the salinity and ionic ratio of the interstitial waters remain constant. For instance, pore waters that move from zones of hydrostatic pressure into zones of geopressures may become undersaturated with respect to carbonates on account of the pressure increase. This could lead to mesogenetic dissolution of carbonate.

Physical processes may create conditions for fracturing and shrinkage. Changes of hydraulic and mechanical stress and also thermal expansion and contraction are common causes for fracturing. The stresses that cause fracturing in sandstones may originate externally, e.g. overburden, or internally within the sandstones, e.g. thermal contraction.

Biochemical processes occasionally trigger void-creating dissolution. A host of boring organisms employ biochemical dissolution of rock material to construct their domiciles or to make space for their holdfasts or roots. Metabolic dissimilation by algae and other organisms can result in carbonic acid and leaching of carbonate. Similarly, sulfate reducing bacteria are capable of causing the dissolution of sulfate minerals.

Biophysical processes cause very little secondary porosity if one discounts the redistribution of primary porosity by burrowing organisms. Mechanical boring and burrowing can create open voids and the growth of roots may lead to fracturing of sandstones.

Mesogenetic decarbonatization porosity. - It is obvious from the preceding discussion that mesogenetic decarbonatization created the vast majority of secondary porosity in sandstones and is, therefore, of the greatest interest. Decarbonatization is here used as a general term for the dissolution of any textural and mineralogical type of carbonate from sandstones or other rocks. The available information suffices to outline various aspects of the occurrence of mesogenetic decarbonatization porosity.

A selection of important occurrences of mesogenetic decarbonatization porosity is listed in Table 2. In all these examples mesogenetic decarbonatization porosity occurs as the predominant or exclusive type of porosity throughout a considerable part of each basin. The incidence of hydrocarbon accumulations in these sandstones indicates that mesogenetic decarbonatization porosity is a very favorable habitat for the trapping of oil and gas. Many giant fields fall into this category including the Prudhoe Bay Field and the Jurassic fields of the North Sea. They represent a significant percentage of global reserves of natural gas and crude oil. Porosity and permeability of these reservoirs can reach very high values (Table 3). Individual samples may have more than 40 volume percent mesogenetic decarbonatization porosity, and permeability of several darcies. Average porosity for

TABLE 2

Significant Occurrences of Mesogenetic Decarbonatization Porosity

AREA	GEOLOGICAL AGE OF SANDSTONES	HYDROCARBON RESERVOIR	SOURCE OF INFORMATION
Mackenzie Delta, Canada	Cretaceous, Tertiary	Yes	Authors
Anderson Plain, Canada	Cambrian	Yes	Authors
Alberta, Canada	Devonian	Yes	Alcock & Benteau
Alberta, Canada	Cretaceous	Yes	Authors
British Columbia Shelf, Canada	Tertiary	No	Authors
Scotian Shelf, Canada	Jurassic, Cretaceous	Yes	Authors
Labrador Shelf, Canada	Cretaceous, Tertiary	Yes	Authors
North Slope, Alaska	Triassic, Jurassic	Yes	Authors
Cook Inlet, Alaska	Jurassic, Cretaceous, Tertiary	Yes	Hayes
Gulf Coast, U.S.A.	Cretaceous, Tertiary	Yes	Stanton & McBride
Wyoming, U.S.A.	Pennsylvanian	Yes	Fox et al
Venezuela	Tertiary	Yes	Authors
North Sea	Permian, Triassic, Jurassic	Yes	Authors
Northern Germany	Pennsylvianian to Jurassic	Yes	Authors
South Africa	Pennsylvanian	No	Rowsell & DeSwardt
Ciscaucasus, Russia	Jurassic, Cretaceous	Yes	Proshlyakov

TABLE 3

Values of Secondary Porosity and Permeability created by Mesogenetic Decarbonatization of Sandstones (based on Laboratory Analyses, Microscopic Analyses, and Log Analyses)

INDIVIDUAL SAMPLES	
POROSITY:	TRACE TO 40 VOL. %
PERMEABILITY:	NIL TO SEVERAL DARCIES
NET PAY ZONES	
POROSITY:	5 VOL. % TO 30 VOL. % (AVERAGE 18 VOL. %)
PERMEABILITY:	A FEW MILLIDARCIES TO 1 DARCY

individual reservoir zones range from 5 to 30 volume percent.

The distribution of the many known examples of mesogenetic decarbonatization porosity can be correlated with a number of geological parameters (Figure 42). The porosity is found in sandstones ranging in geological age from Tertiary to Cambrian and may even occur in Proterozoic sandstones. The ratio of decarbonatization porosity to primary porosity increases with increasing geological age given similar depth and temperature of burial, and similar lithology.

Sandstone lithology places no restriction on the occurrence of mesogenetic decarbonatization porosity. It may be present in sandstones of any primary textural and mineralogical composition in any state of diagenetic alteration. Neither paleogeography nor geological setting appear to have any limiting influence. Mesogenetic decarbonatization porosity occurs in every sizable sedimentary basin in which sandstones have been examined using the criteria discussed earlier in these course notes.

Mesogenetic decarbonatization porosity may originate at depths ranging from shallow to 4,000 meters. Greater depths of origin are probable but have not so far been ascertained. Maximum development of secondary porosity, however, usually occurs within an intermediate depth range, as will be discussed later in this paper. At present, confirmed depths of preservation of mesogenetic decarbonatization porosity range from surface to 6,000 meters. Extrapolation of available data suggests that this porosity type can exist at even greater depths. In sandstones of high mineralogical maturity decarbonatization porosity may possibly extend to a depth of 8,000 meters.

GEOLOGICAL ATTRIBUTES	RANGE OF DECARBONATIZATION POROSITY
GEOLOGIC AGE	CAMBRIAN TO TERTIARY
LITHOLOGY	SANDSTONES OF ANY TEXTURAL AND MINERALOGICAL COMPOSITION
TYPE OF SEDIMENTARY BASIN	ANY SIZABLE SEDIMENTARY BASIN
DEPTH OF ORIGIN OF DECARBONATIZATION POROSITY	SHALLOW TO GREATER THAN 3500M(11,500 FT.)
PRESENT DEPTH	SURFACE TO GREATER THAN 6000M(20,000 FT.)

FIG. 42 - Geologic occurrence of significant mesogenetic decarbonatization porosity in sandstones.

TEXTURAL STAGES OF MESODIAGENESIS

The improved ability to differentiate between primary and secondary porosity permits the tracing of the diagenetic alteration of each of the two porosity classes in response to progressive burial. This provides an excellent means of relating mesogenetic sandstone diagenesis to burial history. We propose four textural stages of sandstone mesodiagenesis that are defined by four petrographically recognizable stages of the burial diagenesis of sandstone porosity (Figure 43). These are in order of progressive burial: (1) immature stage; (2) semi-mature stage; (3) mature stage; and, (4) super-mature stage.

Description. - The immature stage is characterized by mechanical compaction of clean, uncemented sands with primary intergranular porosity (Figure 44). The compaction is accomplished by deformation of ductile grains, rotation of grains, and fracturing of grains. It reduces both primary rock volume, and primary porosity and permeability. Chemical compaction as defined below is negligible or absent. Any significant chemical compaction would tend to lithify the rock and make appreciable mechanical compaction impossible. Secondary porosity may exist. It can either be inherited from eodiagenesis or may originate during the immature stage.

The semi-mature stage is marked by the onset of pervasive chemical compaction, while, at the same time, mechanical compaction of primary porosity becomes insignificant or ceases entirely. We define chemical compaction as being essentially accomplished through dissolution of sand grains at points and interfaces of contact. In the process rock volume and the percentage of intergranular primary porosity are diminished. The dissolved material may reprecipitate within the sandstone

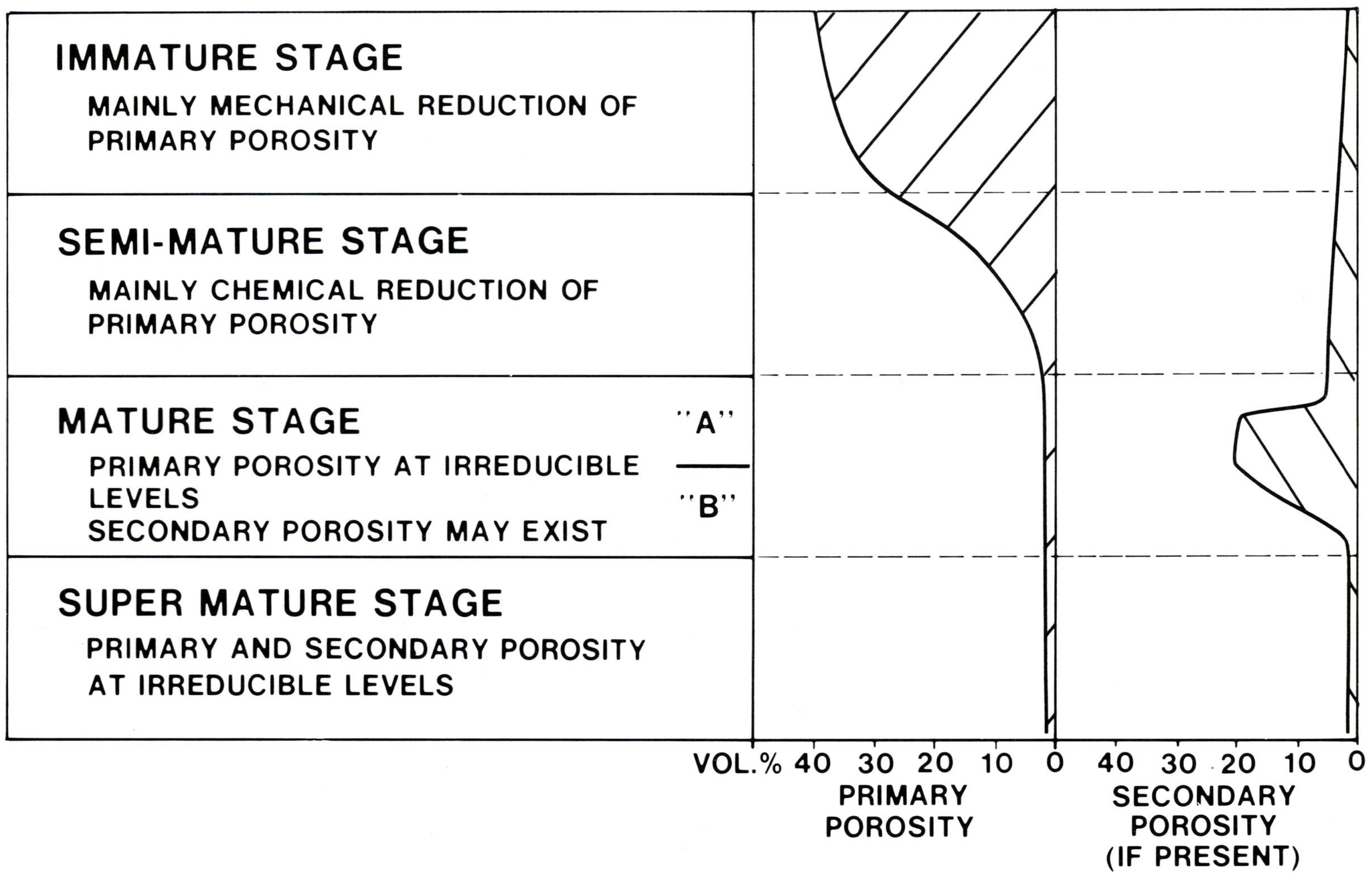

FIG. 43 - Tectural stages of mesodiagenesis of sandstone porosity.

FIGS. 44 to 46 - Stages of immature and semi-mature textural mesodiagenesis

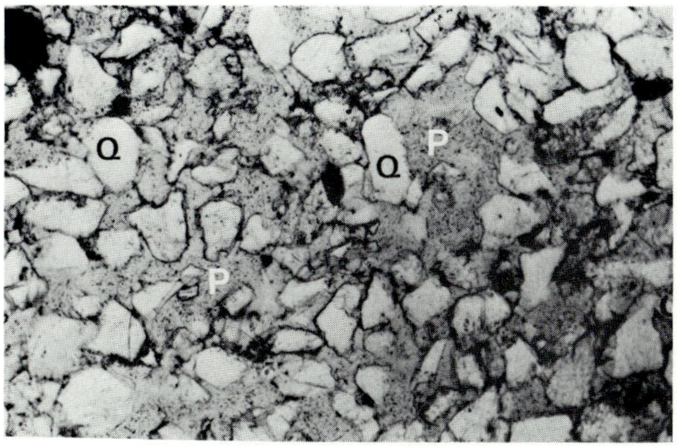

FIG. 44 - Immature stage: Cretaceous quartz arenite, Scotian Shelf, Canada. Polarizer only. P = primary porosity; Q = quartz grains.

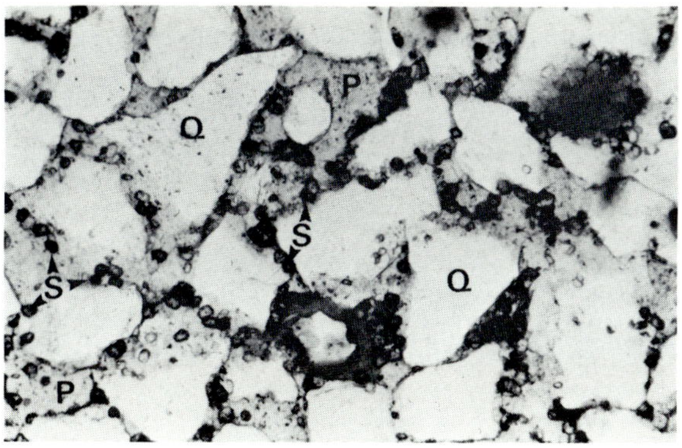

FIG. 45 Semi-mature stage: incipient chemical compaction of Cretaceous quartz arenite, Scotian Shelf, Canada. Polarizer only. P = primary porosity; Q = quartz grains; S = eogenetic siderite cement (very high relief).

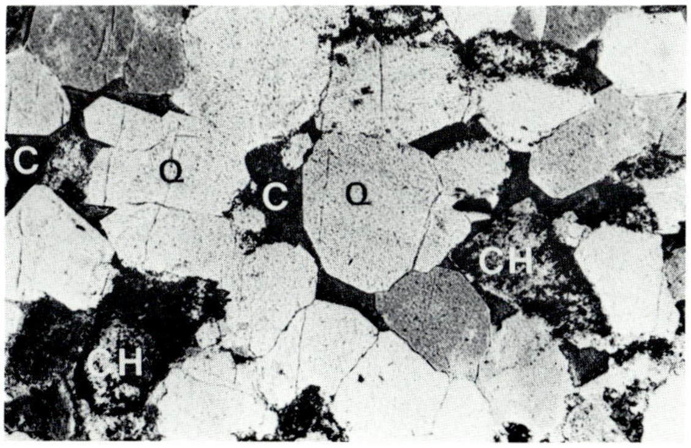

FIG. 46 - Semi-mature stage: advanced cementation of primary porosity by syntaxial quartz overgrowth followed by cementation of residual primary porosity by mesogenetic poikilotopic calcite, Lower Cretaceous quartz arenite, Mackenzie Delta, Canada. Half crossed nicols. Q = quartz grains and mesogenetic quartz overgrowth; CH = chert grains, C = calcite cement.

formation as pore cement further reducing the porosity.

The coexistence of reducible primary porosity and pervasive chemical compaction define the range of the semi-mature stage (Figures 45, 46). Secondary porosity can originate in the semi-mature stage and may coexist with inherited eogenetic and immature stage secondary porosity. At the end of the semi-mature stage the primary porosity has reached irreducible levels. The "irreducible porosity" consists of narrow gaps between grains or diagenetic crystals a fraction of a micron to several microns wide (Figure 47). It is irreducible with respect to chemical compaction and cementation but not to recrystallization. The irreducible porosity in sandstones forms an interconnected system of lamellar pores with extremely low permeability analogous to the irreducible lamellar porosity in carbonate rocks described by Wardlaw (1976). The amount of irreducible porosity decreases with increasing grain size. It commonly ranges from about 0.5 percent to 2.5 volume percent.

The mature stage is characterized by the absence of primary porosity except for irreducible lamellar pores (Figure 48). Secondary porosity can be inherited from preceding stages or may originate during the mature stage (Figure 49). The secondary porosity undergoes chemical compaction and minor mechanical compaction. The mature stage ends when chemical compaction of reducible secondary porosity has reached completion, that is, when it cannot exist any longer for a geologically significant period of time. Two sub-stages can often be identified. Mature stage "A" or early mature stage is the phase of maximum development of secondary porosity. Chemical compaction of secondary porosity is very subordinate during this interval. Mature stage "B" or late mature stage is characterized by the gradual destruction

FIGS. 47 to 49 - Mature stages of textural mesodiagenesis

FIG. 47 - Mature stage: complete loss of reducible primary porosity by chemical compaction and cementation by syntaxial quartz overgrowth, Lower Cretaceous quartz arenite, Mackenzie Delta, Canada. Crossed nicols.

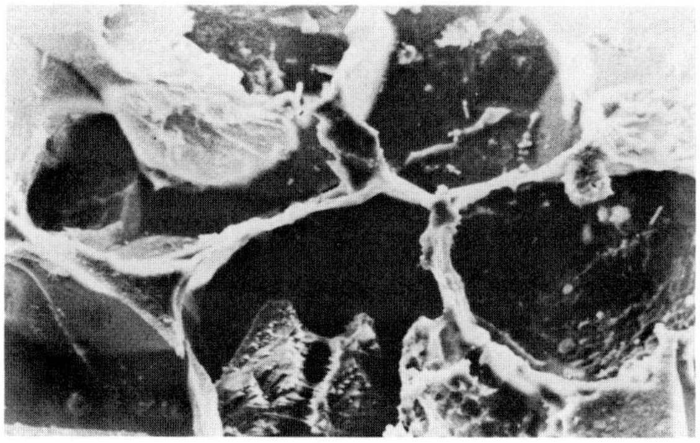

FIG. 48 - Mature stage: epoxy cast of lamellar porosity and secondary intergranular porosity of Lower Cretaceous quartz arenite, Mackenzie Delta, Canada. Scanning-electron micrograph.

FIG. 49 - Mature stage: complete loss of reducible primary porosity by chemical compaction. Subsequent dissolution of sedimentary and diagenetic carbonate created secondary porosity. Jurassic quartz arenite, Beryl Field, North Sea. Half crossed nicols. P = secondary porosity; Q = quartz grains; CH = chert grains; SH = shale laminae.

of secondary porosity mainly by chemical compaction, and little or no generation of secondary porosity.

The supermature stage shows both primary and secondary porosity at irreducible levels. Secondary porosity could conceivably be formed through fracturing and leaching but it would be rapidly eliminated. The advent of metamorphism marks the end of the supermature stage and the conclusion of mesodiagenesis.

In determining the textural stages of mesodiagenesis in sandstones using the above criteria the effects of the following must be discounted: (1) geo-pressures; (2) unusual local chemical conditions; (3) pore filling by hydrocarbons; (4) extrastratal cementation; and (5) unusual local conditions of temperature and pressure such as might be caused by tectonism and vulcanism. The petrographic analysis of a sandstone cannot, by itself, determine whether it presently still resides within the zone of mesodiagenesis that left the final textural imprint. This question must be answered by geological analysis of the sedimentary basin. It is possible to deduce from the present distribution of regimes of mesodiagenesis the zones of currently active burial diagenesis for specific sandstone lithologies. In similar fashion, one is able to reconstruct ancient regimes of mesodiagenesis.

Influence of sandstone mineralogy. - Figure 50 shows a schematic plot of the depth ranges of textural maturity grades of burial diagenesis versus mineralogical maturity of sandstones. The basic assumptions for the plot are: (1) the subsidence rate of the sedimentary basin is thirty meters per one million years; (2) the sedimentation rate equals the sum of subsidence and compaction; (3) the geothermal gradient is 2.7 degrees Celsius per one hundred meters; and (4) the

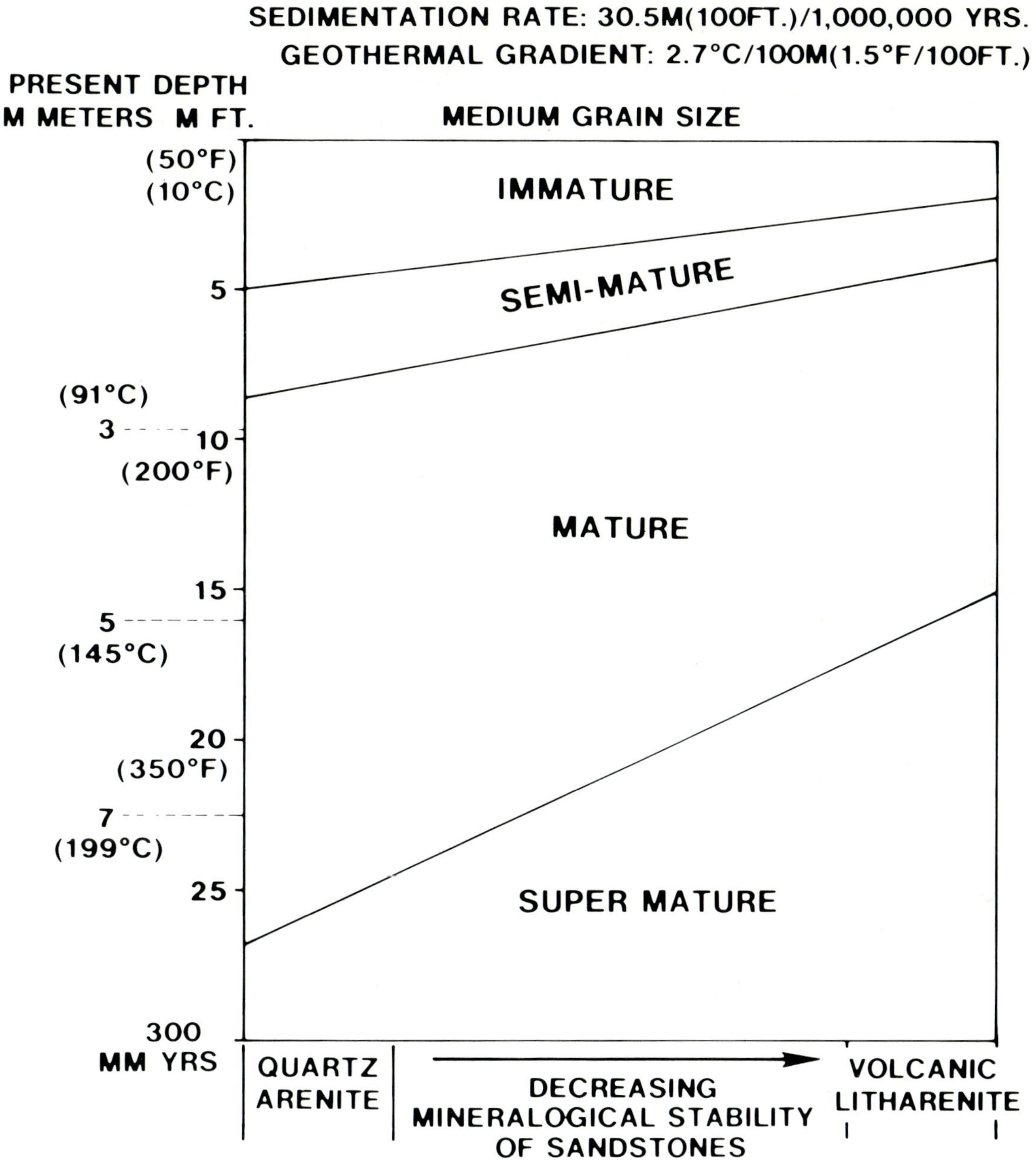

FIG. 50 - Influence of sandstone mineralogy on burial diagenesis of porosity.

sandstones are medium grained. Depths of individual zones of textural diagenetic maturity become shallower with decreasing mineralogical maturity, a fact which has been observed by many workers (Galloway, 1974).

Surprisingly, the transition from the semi-mature zone to the mature zone occurs at relatively shallow depths. This implies that in most sedimentary basins effective primary sandstone porosity is restricted to less than half the sedimentary column. The volume of the mature zone in the majority of sedimentary basins exceeds the combined volumes of the immature and semi-mature zones. Thus the volume of sediments which is prospective for secondary porosity is more than double the volume of sediments which is prospective for effective primary porosity.

The large vertical extent of the mature zone compared with that of the semi-mature zone is unexpected. The higher temperature of the mature zone must increase the rate of reaction of chemical compaction if other conditions remain constant. It is obvious that some of these other conditions must have changed thereby hindering chemical compaction. At the present time it is not certain which changes are responsible for the delay in the destruction of secondary porosity in the mature zone. Future investigations will have to solve this problem. It is possible that changes in the chemical environment or changes in the pore geometry and/or changes in the submicroscopic morphology of the interfaces at grain contacts may be the cause.

<u>Influence of temperature and time</u>. - The active stages of burial diagenesis of porosity are schematically plotted in Figure 51 in relationship to residence temperature (residence depth) and residence time. The assumed sandstone lithology is

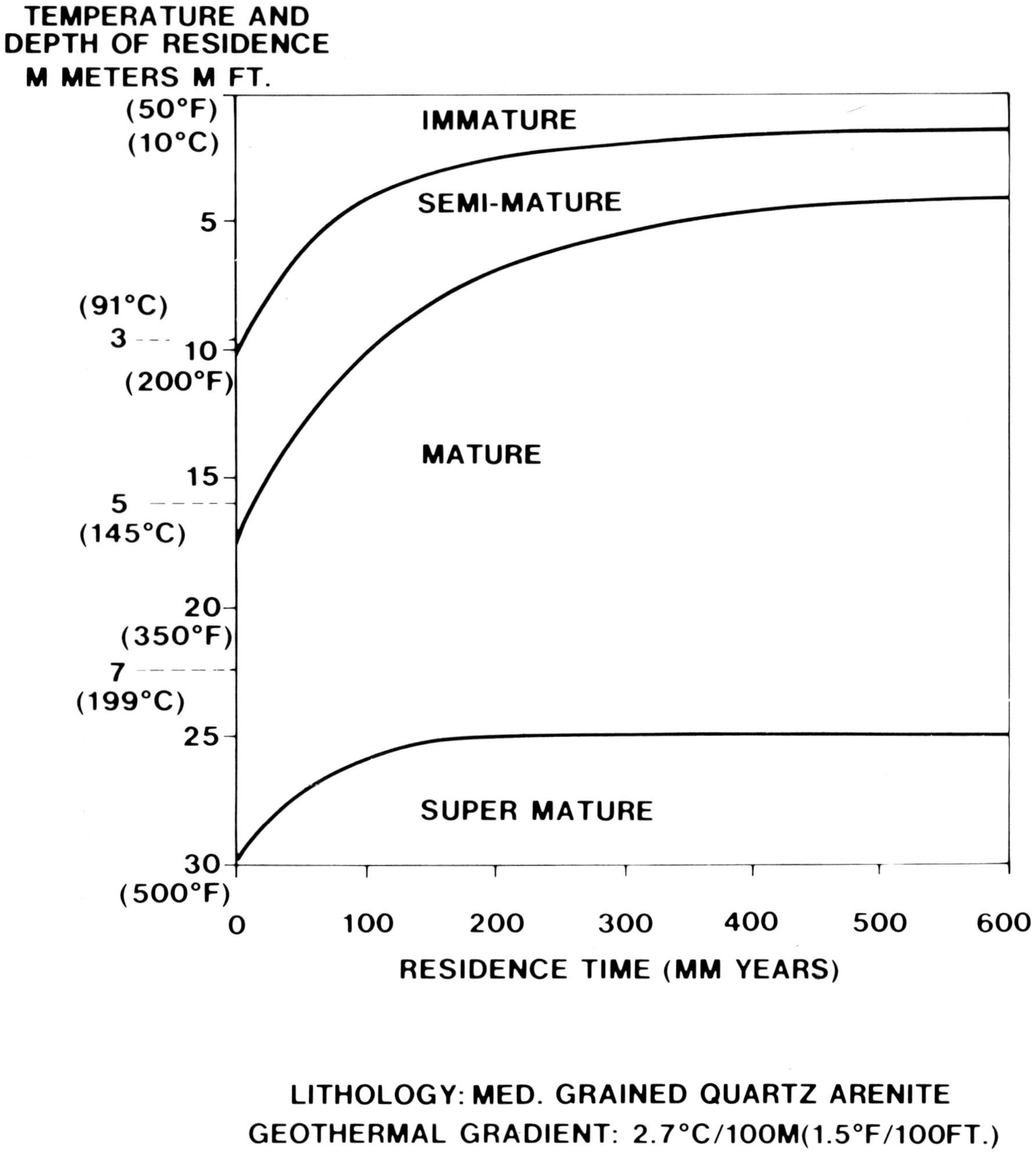

FIG. 51 - Influence of temperature and time on burial diagenesis of porosity.

medium grained quartz arenite and the chosen geothermal gradient is 2.7 degrees Celsius per one hundred meters. It can be seen that the level of maturity increases with both increasing residence temperature and time. Lengthening of the residence time beyond 400 million years, however, produces very little change.

A similar relationship would be found if the diagram were plotted for other mineralogical classes of sandstone. The position of the zones of textural maturity would shift somewhat towards the upper left (lower residence temperature and shorter residence time) with decreasing mineralogical maturity. These relationships indicate that the texturally mature zone becomes volumetrically more important as the geothermal gradient and the age of the sediments increase.

Porosity range. - Figure 52 shows a schematic plot of the range of primary and secondary porosity versus depth in medium grained quartz arenites. The assumptions made in constructing the plot are the same as for Figure 50.

Maximum primary porosities decline with increasing depth as a result of mechanical and chemical compaction. At a depth of about 3,000 meters the mature zone is reached and only one to two volume percent of irreducible primary porosity remains.

Maximum secondary porosities exceed 40 volume percent above 4,000 meters, that is, in the immature, semi-mature and upper part of the mature zone. From there they decline steadily with increasing depth in the middle and lower part of the mature zone. At an extrapolated depth of about 8,000 meters secondary porosity should diminish to irreducible levels.

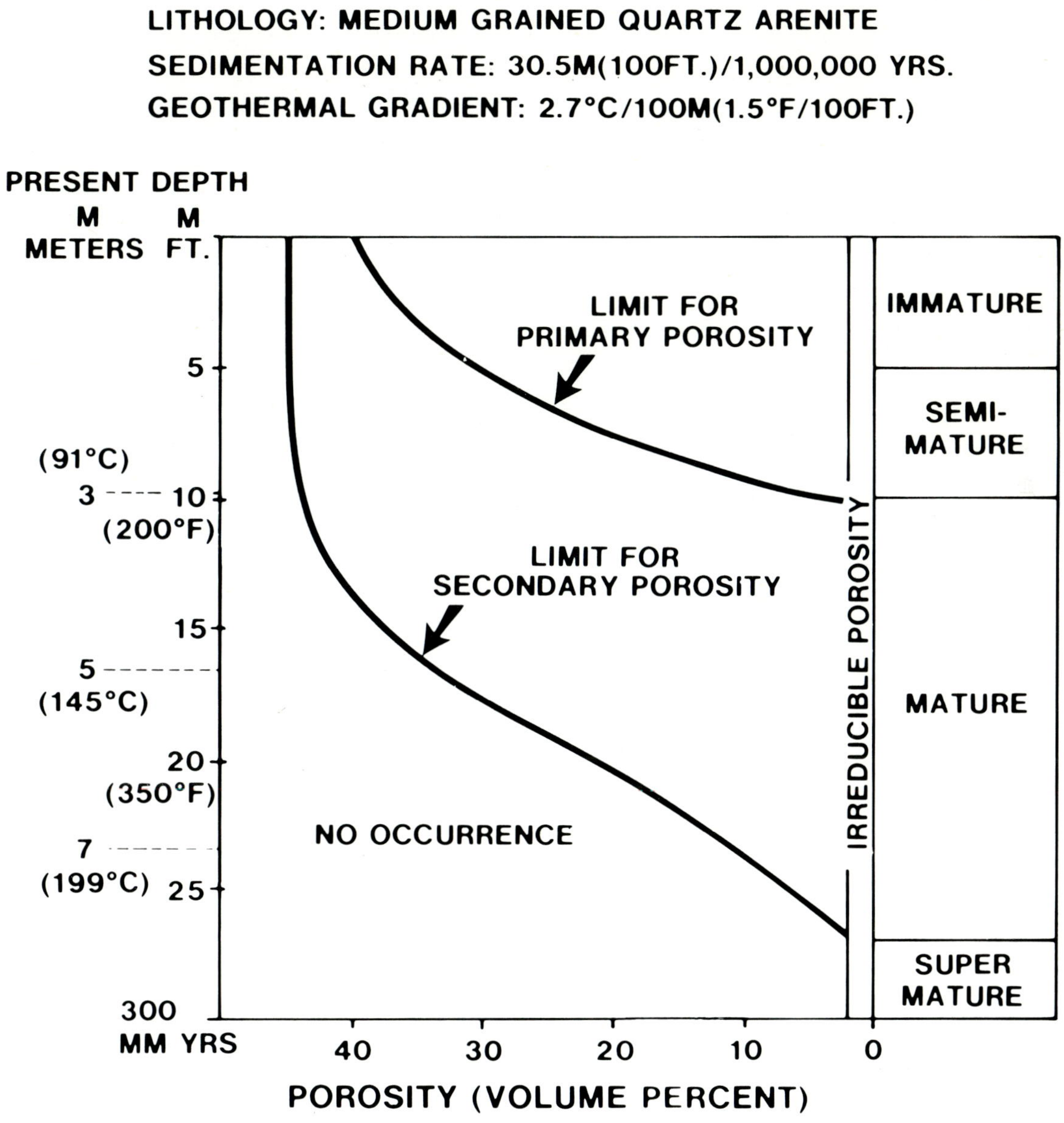

FIG. 52 - Maximum porosity of quartz arenites versus depth of burial.

Significantly, secondary porosity can be higher than original primary porosity in the immature, semi-mature, and upper mature zones. In addition, the depth limit for porosity over 10 volume percent is more than twice as deep in the case of secondary porosity compared to that of primary porosity. The mineralogically less mature sandstones exhibit qualitatively similar relationships between the textural stages of mesodiagenesis and porosity (Galloway, 1974; Hayes, 1979).

DIAGENESIS OF QUARTZ ARENITES

The sandstones of the highest mineralogical stability, the quartz arenites, have been selected to trace in detail the role of secondary porosity during the course of their diagenesis. This choice was made for two reasons: (1) the data base for this sandstone lithology is the most comprehensive; and (2) primary porosity survives much longer in quartz arenites than in sandstones of lower mineralogical maturity. Secondary porosity may be expected to play a relatively less important role in quartz arenites compared to the latter. Thus a baseline for the significance of secondary porosity can be established.

It is necessary to follow the path of typical diagenetic histories of individual sandstones in order to better understand the vertical distribution of primary and secondary porosity. Four types of diagenetic histories will be discussed: (1) quartz arenites free of carbonate; (2) quartz arenites with sedimentary and eogenetic carbonate; (3) quartz arenites with mesogenetic carbonate; (4) quartz arenites with sedimentary, eogenetic and mesogenetic carbonate.

The diagenesis diagrams (Figures 53, 54, 55, 59, 63, 67)

chart the composition of hypothetical quartz arenites from the time of deposition to the advent of metamorphism. These diagrams have been constructed by piecing together petrographical and geological observations of numerous samples.

Quartz arenites free of carbonate. - The loss of primary porosity in quartz arenites is best examined in samples that remain free of carbonate or other readily soluble constituents throughout their diagenesis. The primary intergranular porosity and the rock volume of the quartz arenite shown in Figure 53 are reduced by mechanical compaction during the immature stage. Chemical compaction and reprecipitation of the dissolved silica as cement reduce the intergranular porosity to irreducible lamellar porosity during the semi-mature stage. A considerable reduction in rock volume accompanies this process. Thereafter the sandstone remains unchanged until some recrystallization occurs during the supermature stage, (Figure 56).

The quartz arenite illustrated in Figure 54 received silica cement from an extrastratal source with the onset of the semi-mature stage. The silica cementation prevented chemical compaction (Figure 57). Many coarse grained quartz arenites show this type of diagenesis. Possible sources of the extrastratal silica cement include: (1) chemical compaction of adjacent very fine grained quartz arenites and quartz siltstone; (2) replacement of quartz grains by carbonate; (3) degradation of feldspars; and (4) clay mineral diagenesis in adjacent shales.

The quartz arenite shown in Figure 55 experienced chemical compaction without reprecipitation of the dissolved silica as cement. Therefore, the rock volume is greatly reduced during the semi-mature stage (Figure 58). Of necessity, the dissolved

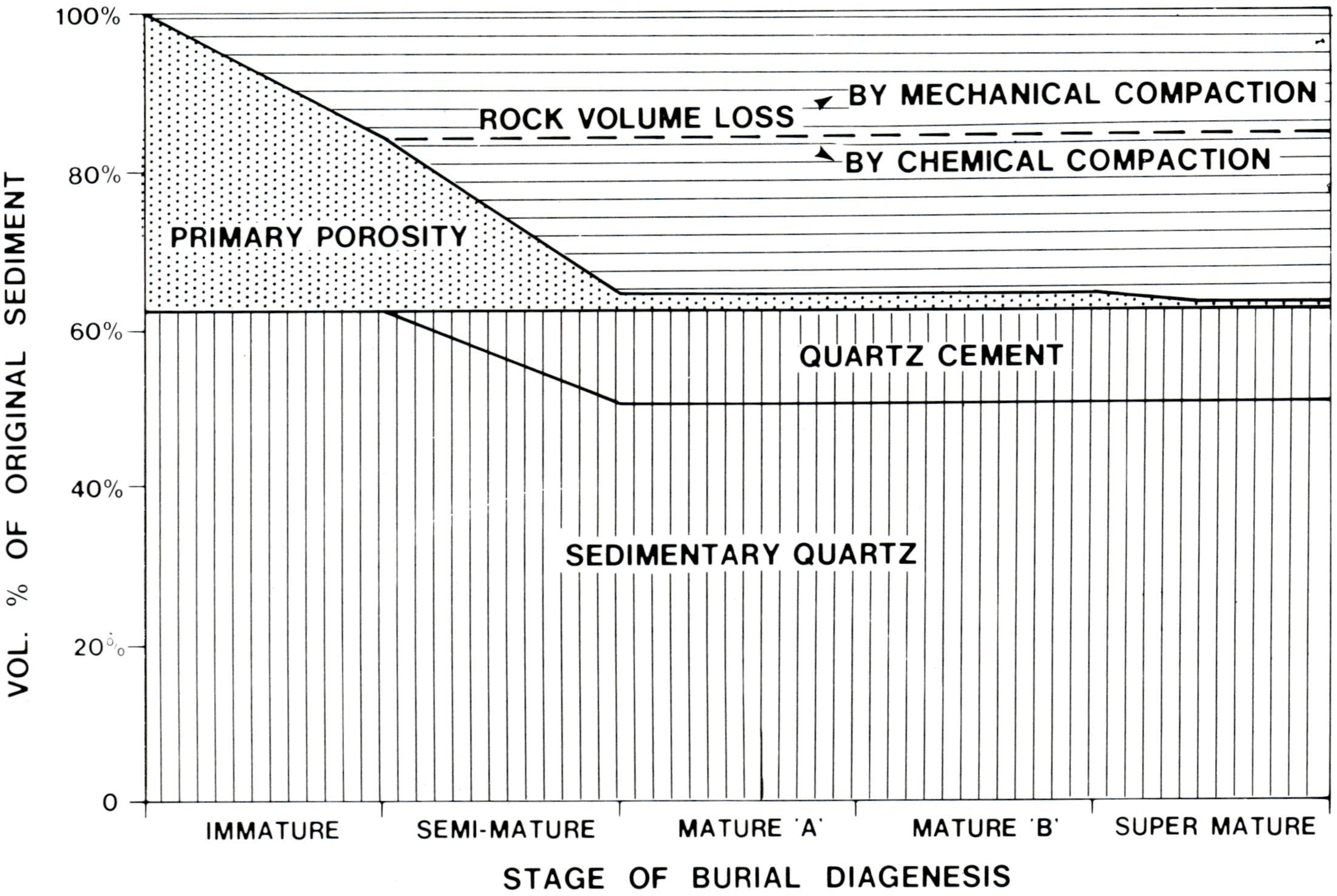

FIG. 53 - Burial diagenesis of quartz arenite with quartz dissolution equal to quartz precipitation.

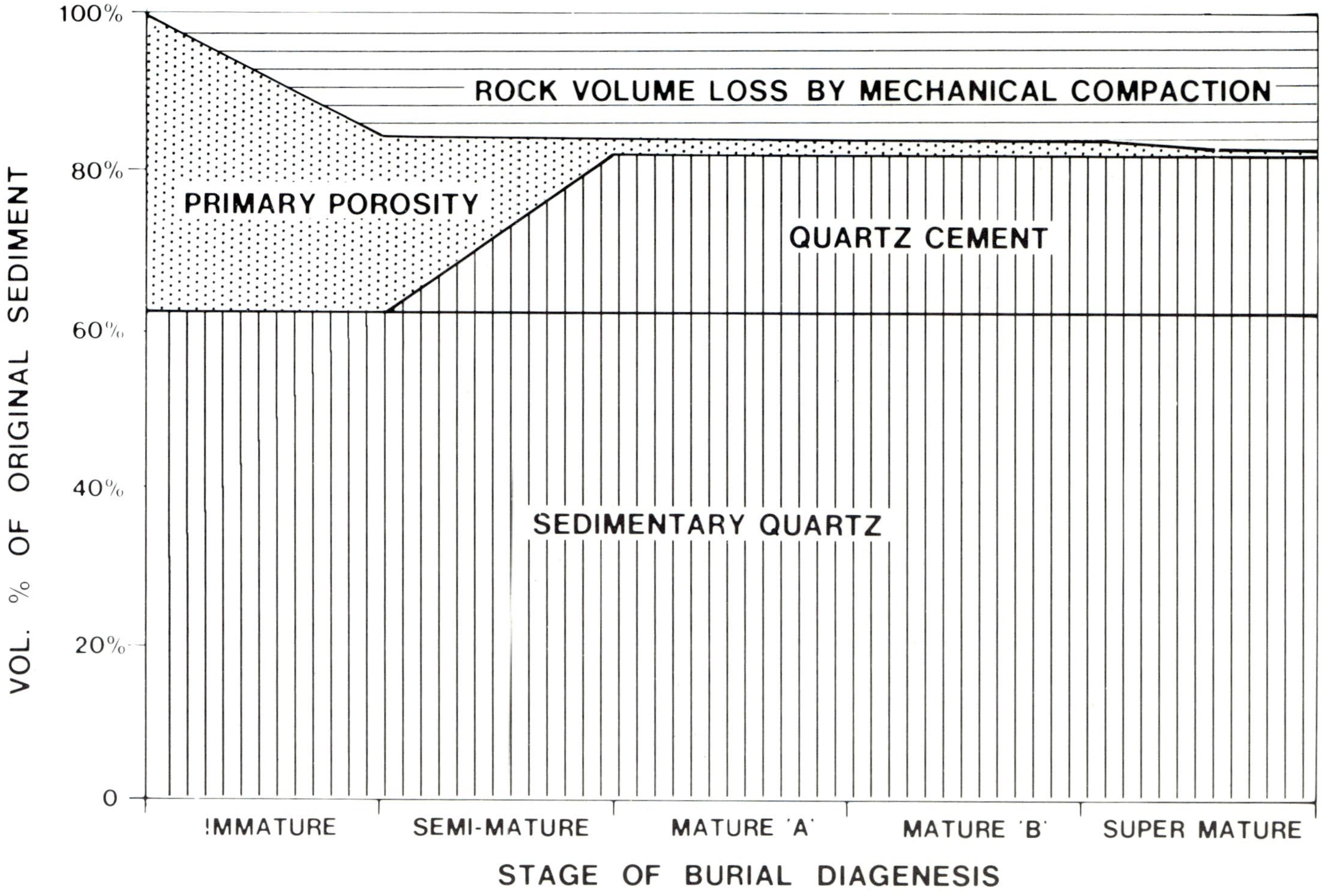

FIG. 54 - Burial diagenesis of a quartz arenite with extrastratal source of quartz cement and without quartz dissolution at grain contacts.

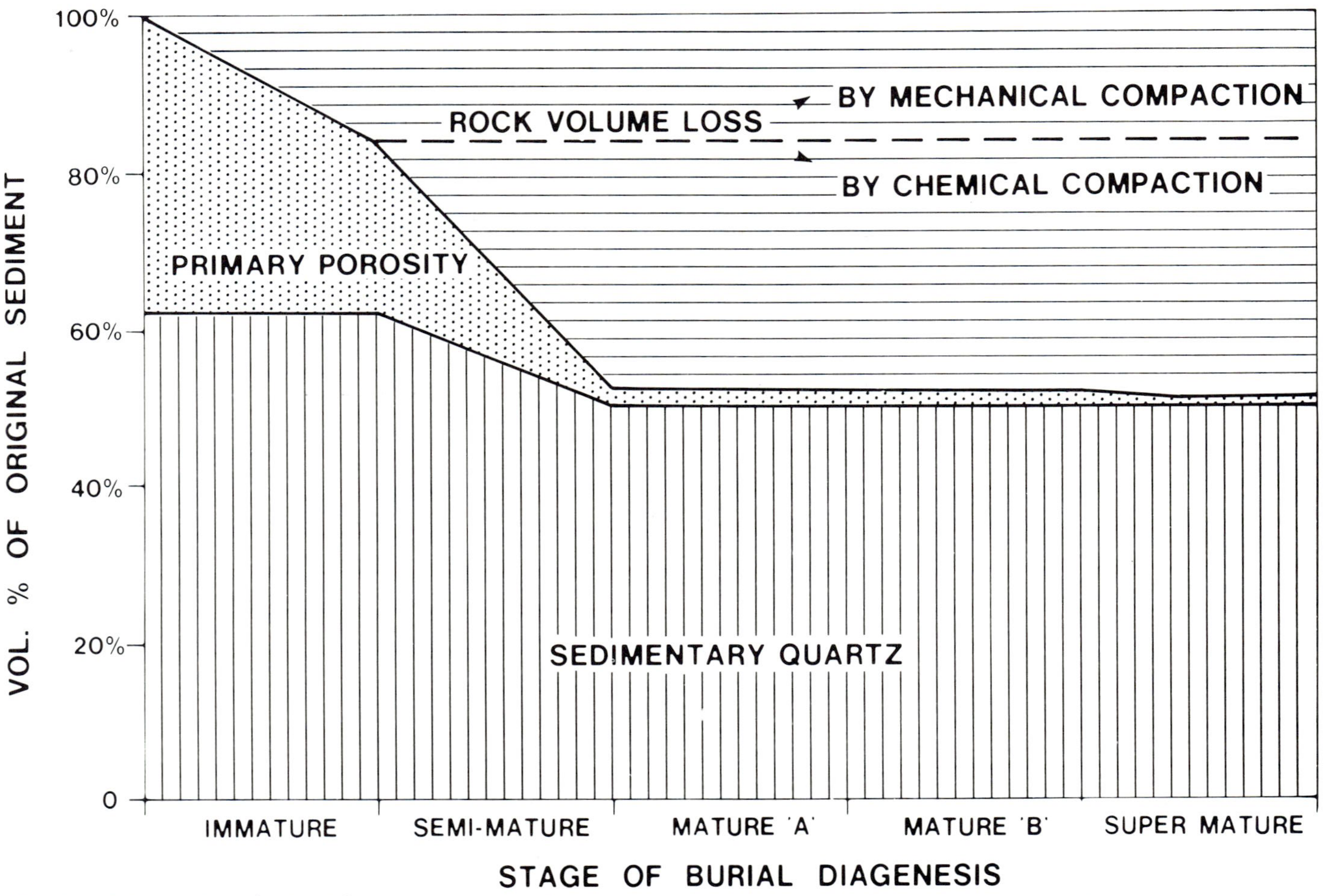

FIG. 55 - Burial diagenesis of quartz arenite with quartz dissolution at grain contacts and without cementation.

FIGS. 56 to 58 - Porosity diagenesis of quartz arenite lacking carbonate

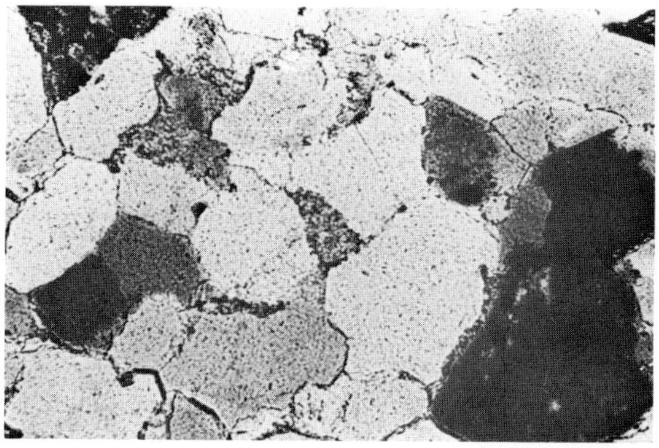

FIG. 56 - Complete loss of reducible primary porosity by quartz dissolution along grain contacts (chemical compaction) and syntaxial quartz overgrowth. Quartz arenite lacking carbonate. Mature stage. Lower Cretaceous, Mackenzie Delta, Canada. Half crossed nicols.

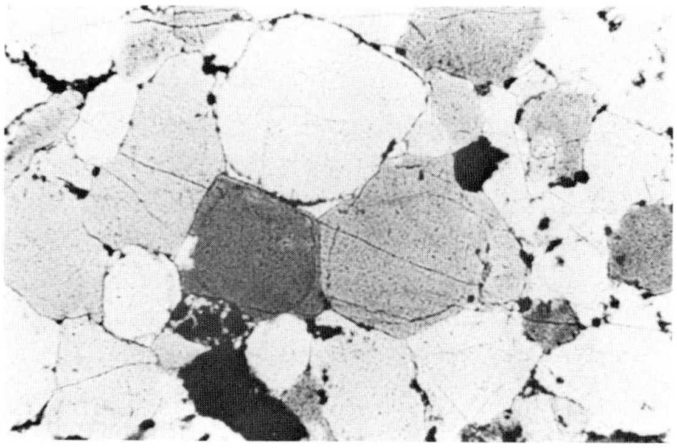

FIG. 57 - Complete loss of reducible primary porosity due to cementation by syntaxial quartz overgrowth (chemical compaction). Quartz arenite lacking carbonate. Mature stage. Lower Cretaceous, Mackenzie Delta, Canada. Half crossed nicols.

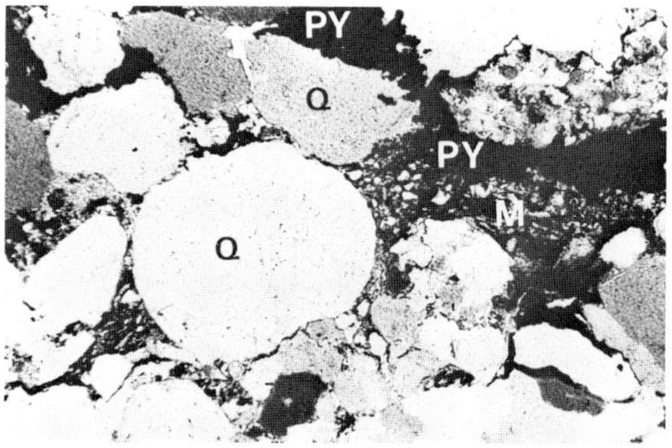

FIG. 58 - Complete loss of reducible primary porosity by quartz dissolution at grain contacts (chemical compaction). Quartz arenite, silty/argillaceous, lacking carbonate. Mature stage. Lower Cretaceous, Mackenzie Delta, Canada. Half crossed nicols. Q = quartz grains; M = silty/argillaceous matrix; PY = pyritized matrix.

silica migrates away from the sandstone bed. Very fine grained quartz arenites frequently follow this diagenetic path.

The three case histories outlined above demonstrate the early loss of primary porosity. Clearly, mesogenetic decarbonatization porosity cannot develop because of the absence of carbonate throughout the history of the rocks. Such carbonate-free quartz arenites are best suited for determining the first three stages of burial diagenesis.

Quartz arenites with sedimentary and/or eogenetic carbonate. - A large percentage of quartz arenites contain significant amounts of sedimentary and eogenetic carbonate (as also do other sandstone lithologies). The sources of this carbonate are: (1) erosion of carbonate rocks; (2) marine environments; (3) evaporitic environments; (4) deltaic environments; (5) lacustrine and paludal environments; (6) fluviatile environments; and, (7) subaerial environments. The dominant original carbonate minerals are siderite, dolomite, low-magnesium calcite, high magnesium calcite, and aragonite. In most instances the latter two minerals subsequently convert to low-magnesium calcite.

The diagenesis diagram of a fairly common type of quartz arenite is shown in Figure 59. The original sand contains some sedimentary carbonate. Carbonate cement fills part of the intergranular porosity in the eogenetic regime and, in doing so, moderately lithifies the sandstone. Because of this lithification, no mechanical compaction occurs during the immature stage of mesodiagenesis. Chemical compaction takes place during the semi-mature stage, considerably reducing rock volume and porosity. The dissolved silica reprecipitates as intergranular cement within the sandstone bed and further reduces the porosity. At the beginning of the mature stage

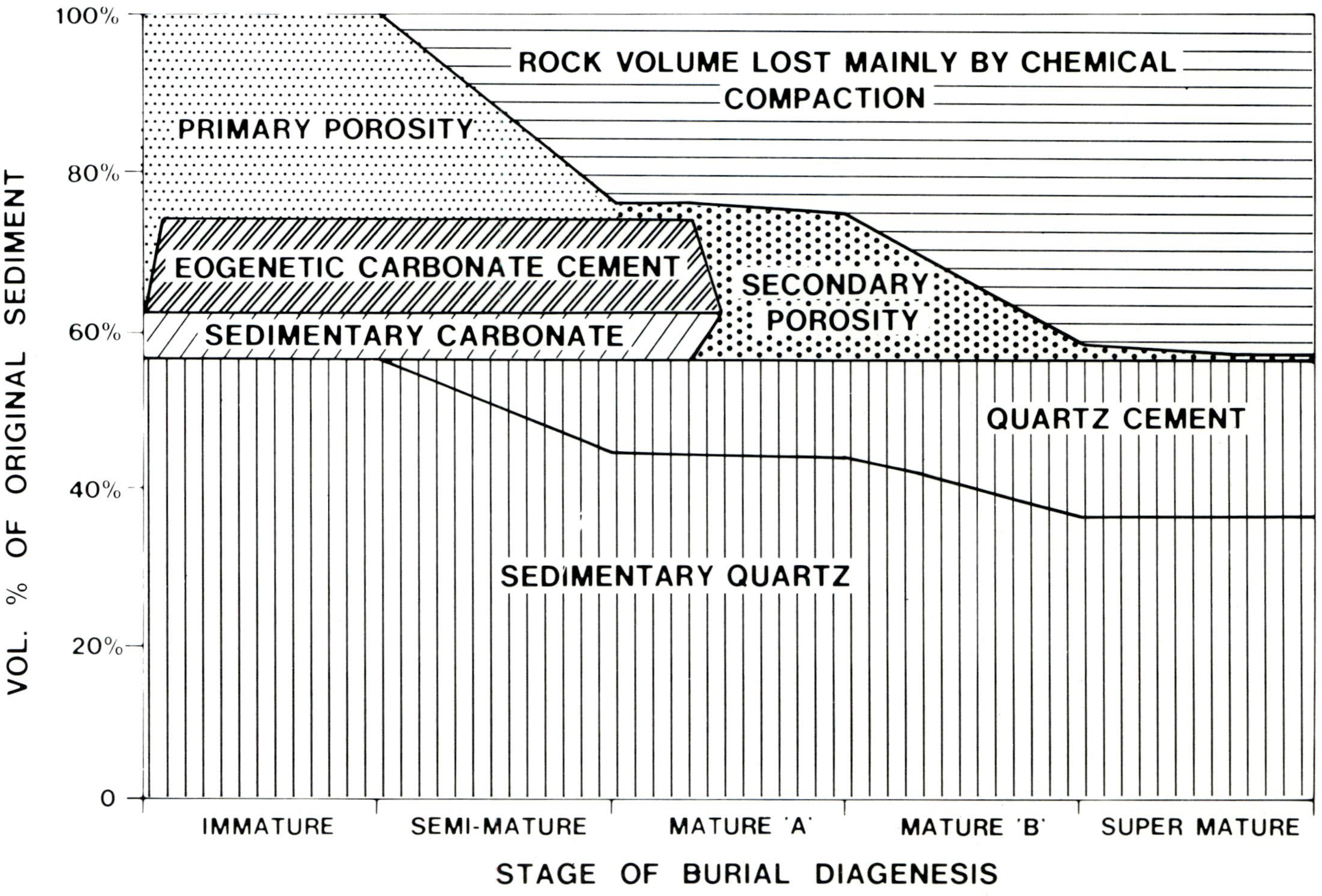

FIG. 59 - Burial diagenesis of a quartz arenite showing eogenetic carbonatization and mesogenetic decarbonatization.

"A", the sandstone contains only irreducible lamellar porosity (Figure 60). However, a considerable amount of porosity is subsequently created through decarbonatization (Figures 61 & 62). Minor mechanical and chemical compaction occurs after complete decarbonatization during the remaining mature stage "A". Chemical compaction becomes more active during the mature stage "B" and the secondary porosity declines to an irreducible remnant of lamellar porosity. At this point the supermature stage is reached.

This diagenetic history demonstrates that secondary porosity survives chemical compaction longer during deep burial than primary porosity does at much shallower depth. Decarbonatization can occur at any stage of burial diagenesis and individual sandstones may be subjected to one or several periods of decarbonatization. However, decarbonatization reaches a pronounced maximum at the beginning of the mature stage "A" in quartz arenites. Decarbonatization is certainly not inevitable; many quartz arenites containing carbonates are not affected by it and remain tight throughout the semi-mature and mature stages.

Quartz arenites with mesogenetic carbonate. - Many quartz arenites that were buried containing no carbonate, later receive mesogenetic carbonate in the form of cement or replacement. The dominant mesogenetic carbonate minerals are calcite, dolomite, and siderite. Mesogenetic carbonatization is frequently succeeded by partial or complete decarbonatization.

Figure 63 illustrates a typical case history of mesogenetic carbonatization and decarbonatization. Diagenesis during the immature and early semi-mature stages follows the same path as in Figure 53. Carbonatization during the semi-

FIGS. 60 to 62 - Porosity diagenesis of quartz arenites with sedimentary and eogenetic carbonate

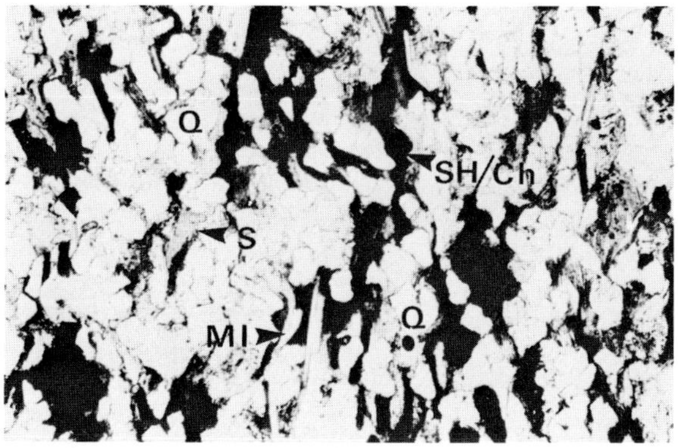

FIG. 60 - Complete loss of reducible primary porosity due to mechanical compaction and dissolution at grain contacts. Quartzose/sideritic/argillaceous/micaceous litharenite with abundant sedimentary and eogenetic siderite. Mature stage. Jurassic, Statfjord Field, North Sea. Polarizer only. Q = quartz grains; S = siderite; MI = mica; SH/Ch = shale fragments and chert grains.

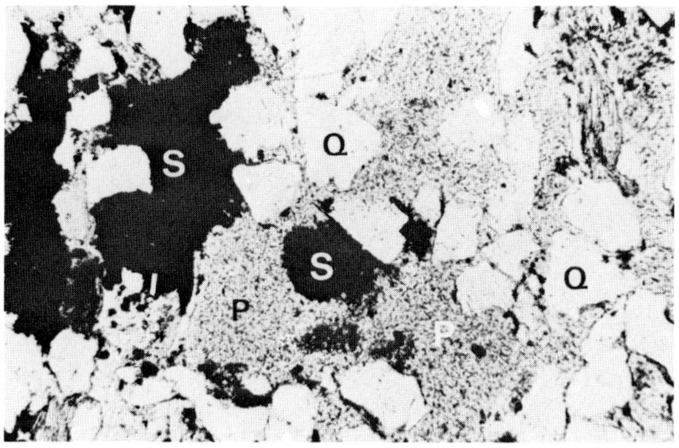

FIG. 61 - Partial mesogenetic dissolution of sedimentary and eogenetic siderite created secondary porosity. Sideritic/quartzose litharenite. Mature stage. Jurassic, Statfjord Field, North Sea. Polarizer only. P = secondary porosity; Q = quartz grains; S = siderite.

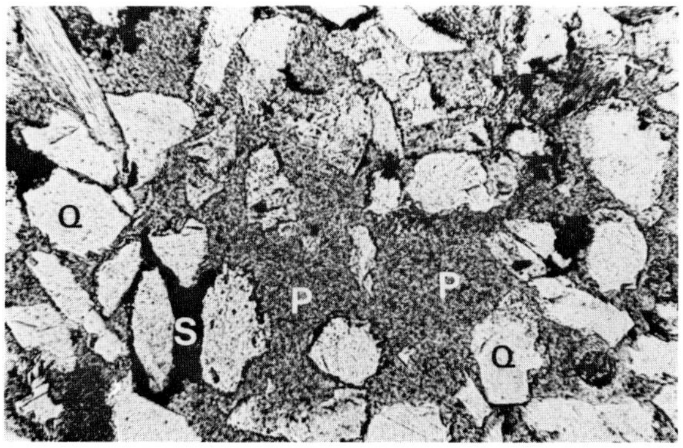

FIG. 62 - Near complete mesogenetic dissolution of sedimentary and eogenetic carbonate created 40 volume percent secondary porosity (core analysis). Same original composition as Fig. 61. Mature stage. Jurassic, Statfjord Field, North Sea. Polarizer only. P = secondary porosity; Q = quartz grains; S = siderite.

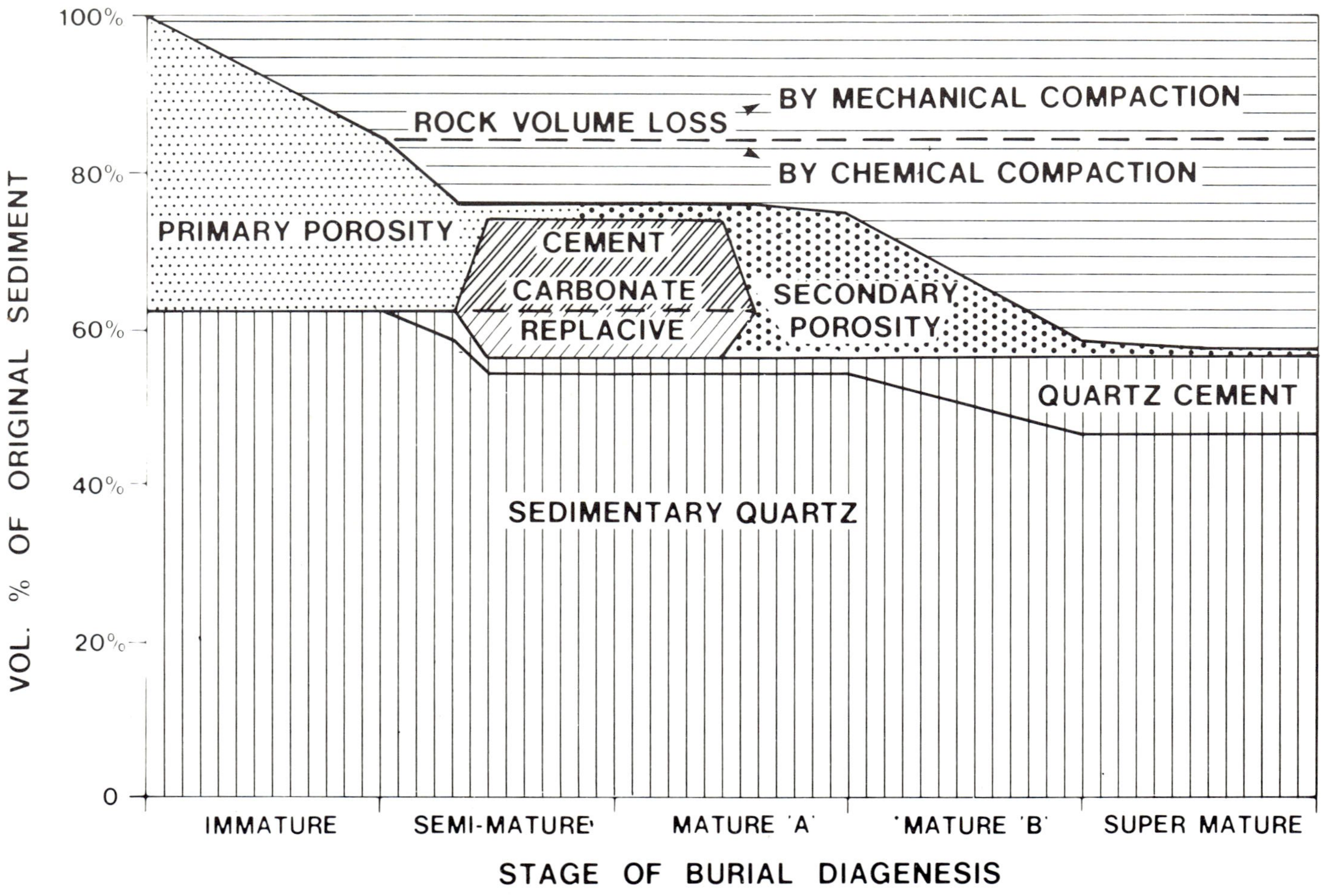

FIG. 63 - Burial diagenesis of a quartz arenite with mesogenetic carbonatization and decarbonatization.

mature stage reduces the remaining intergranular porosity to an irreducible volume of lamellar porosity and replaces part of the quartz cement and margins of quartz grains (Figure 64). The carbonatization also halts chemical compaction. No further change occurs until decarbonatization takes place during the early mature stage (Figures 65 & 66). The volume of the decarbonatization porosity is larger than the volume of primary porosity at the time of carbonatization because carbonate cement and replacive carbonate have both been completely leached. The subsequent diagenesis follows a similar pattern to that described for Figure 59.

Carbonatization can occur at any stage of mesogenetic sandstone diagenesis. However, in quartz arenites a distinct maximum of carbonatization occurs during the last two-thirds of the semi-mature stage with carbonate cement occupying mainly primary porosity. This zone of maximum carbonatization in many sedimentary basins forms a distinct zone of carbonate enrichment which may be referred to as a "carbonate curtain."

During the process of mesogenetic carbonatization of quartz arenites large volumes of quartz are replaced by carbonate. The silica must go into aqueous solution and is available for reprecipitation. This silica appears to be a major source of the silica overgrowth cement which reduces primary porosity during the semi-mature stage.

Quartz arenites with sedimentary, eogenetic and mesogenetic carbonate. - A fairly common, although complex, case history of diagenesis is shown in Figure 67. The original quartz sand contains a moderate amount of sedimentary carbonate. Eogenetic replacement and cementation add some more carbonate and slightly consolidate the sediment. Mechanical compaction during the immature stage is only minor because of

FIGS. 64 to 66 - Porosity diagenesis of quartz arenites

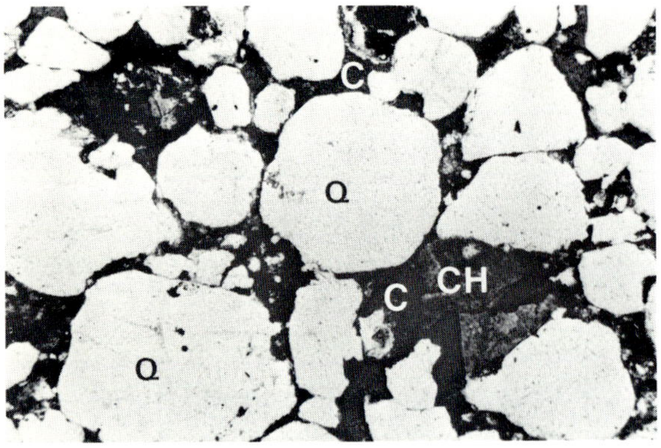

FIG. 64 - Complete loss of reducible primary porosity due to (1) mechanical compaction (fractured chert grain); (2) incipient chemical compaction; and, (3) cementation by mesogenetic poikilotopic calcite. Quartz arenite. Mature zone. Lower Cretaceous, Mackenzie Delta, Canada. Polarizer only. Q = quartz grains; CH = chert grain; C = calcite cement.

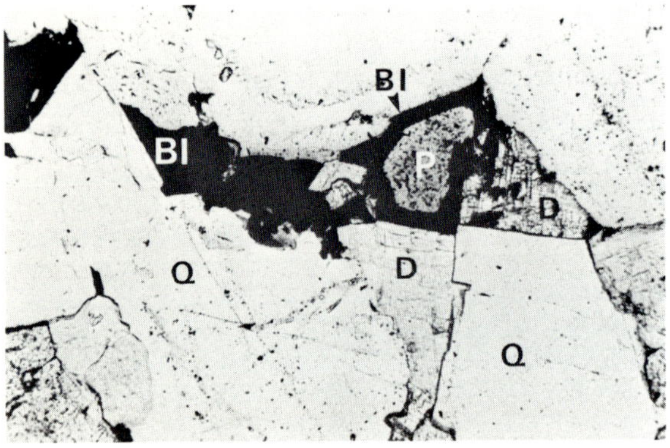

FIG. 65 - Partial mesogenetic dissolution of mesogenetic dolomite created secondary porosity. Mesogenetic dolomite replaced quartz overgrowths and completely cemented reducible primary porosity. Migration of crude oil occurred after decarbonatization as documented by bitumen coating secondary pores. Mature zone. Cambrian, Northwest Territories, Canada. Polarizer only. P = secondary porosity; Q = quartz grains and syntaxial quartz overgrowth; D = cementing and replacive dolomite; BI = bitumen.

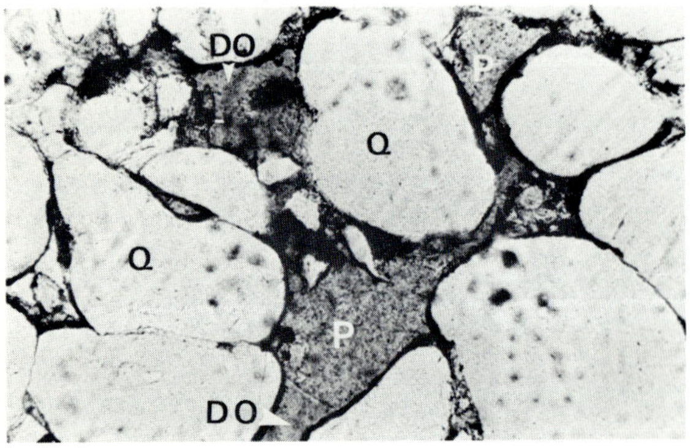

FIG. 66 - Complete dissolution of mesogenetic calcite cement created secondary porosity. Migration of crude oil occurred after decementation and left secondary pores partially filled with dead oil. Mature zone. Cambrian, Northwest Territories, Canada. Polarizer only. P = secondary porosity; Q = quartz; DO = dead oil.

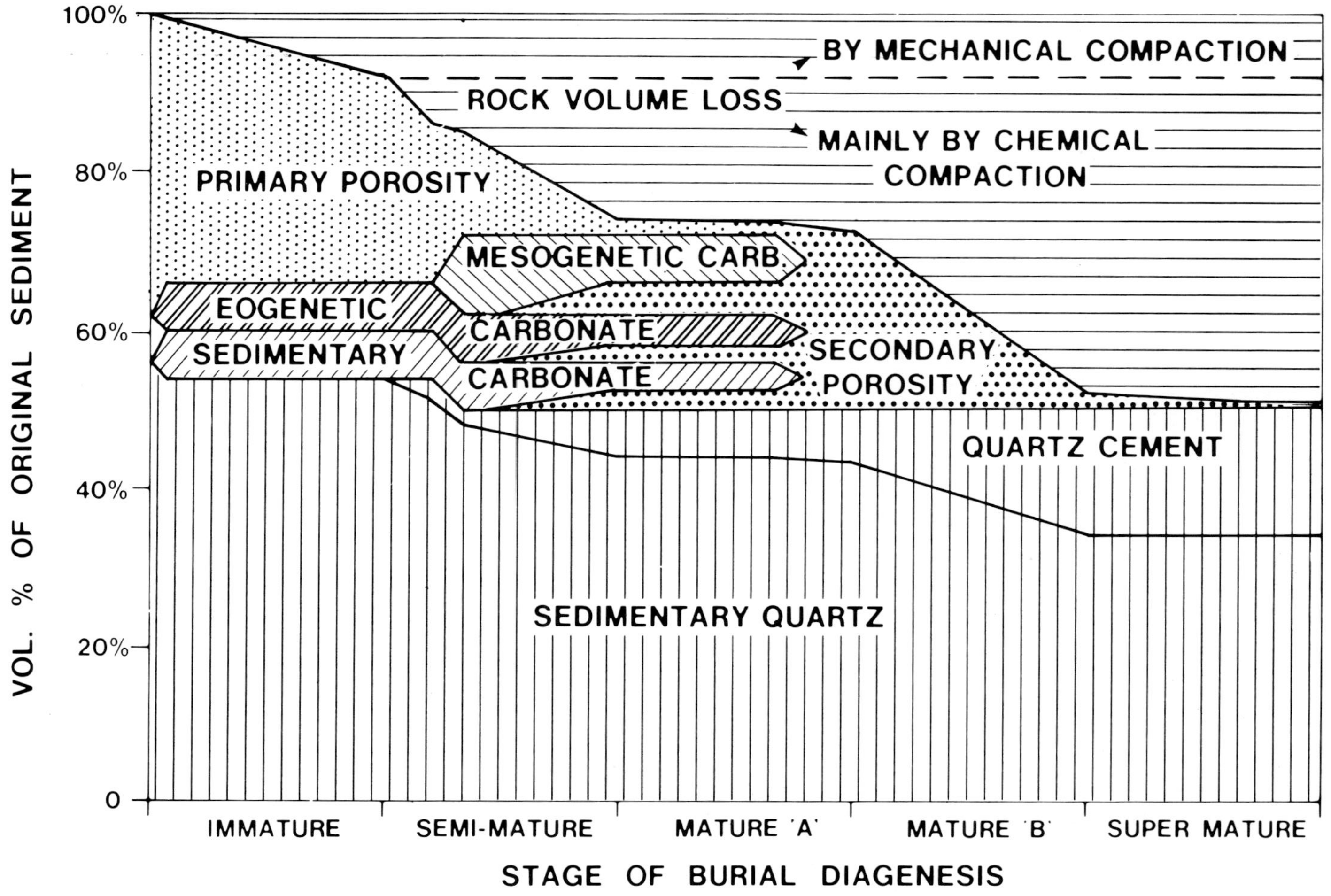

FIG. 67 - Burial diagenesis of a quartz arenite showing eogenetic carbonatization and mesogenetic carbonatization with mesogenetic decarbonatization creating an overlap.

the eogenetic carbonatization. Some mesogenetic carbonatization, including both replacement and cementation, occurs after some chemical compaction during the semi-mature stage. Chemical compaction resumes its activity during the remaining part of the semi-mature stage because mesogenetic cementation is incomplete (Figure 68). At the same time incomplete decarbonatization takes place creating some secondary porosity. The sandstone enters the mature zone containing considerable amounts of secondary porosity and a small amount of irreducible primary porosity (Figure 69). During mature stage "A" all remaining carbonate constituents are dissolved, greatly increasing the amount of secondary porosity (Figure 70). Destruction of the secondary porosity during mature stage "B" follows a similar pattern to that described for Figure 59. The final rock volume in the supermature stage is only half that of the original quartz sand.

The above case history shows that decarbonatization can create exceedingly high porosities in quartz arenites which contain sedimentary, eogenetic, and mesogenetic carbonate including a considerable fraction of replacement carbonate. Multiple pulses of incomplete decarbonatization are very common, but are extremely difficult to recognize and differentiate microscopically. Individual quartz arenites may also be subjected to several phases of mesogenetic carbonatization, some of which can be readily distinguished under the petrographic microscope. Usually, electron luminescence microscopy and microprobe analysis must be applied to solve the more complex sequences of diagenetic events.

SYNTHESIS. - The principal observations concerning the burial diagenesis of quartz arenites are summarized in Figure 71. The sequence of diagenetic events is mainly controlled by residence temperature and residence time.

FIGS. 68 to 70 - Porosity diagenesis of quartz arenites

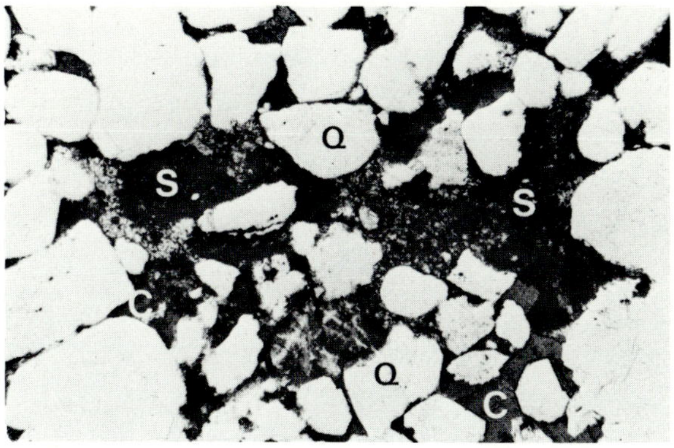

FIG. 68 - Complete loss of reducible primary porosity in sideritic quartz arenite. Sequence of porosity loss: (1) mechanical compaction of siderite pellets; (2) chemical compaction; and, (3) mesogenetic cementation by poikilotopic calcite. Mature zone. Lower Cretaceous, Mackenzie Delta, Canada. Polarizer only. Q = quartz; S = siderite; C = calcite.

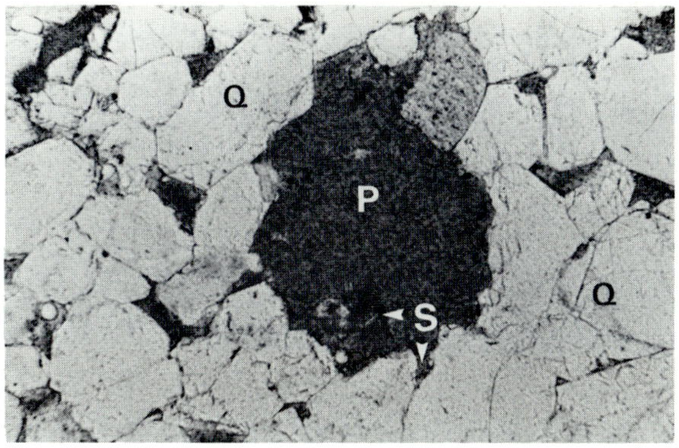

FIG. 69 - Near complete dissolution of sedimentary and eogenetic siderite and mesogenetic calcite created secondary porosity in quartz arenite, following after complete loss of reducible primary porosity. Mature zone. Lower Cretaceous, Mackenzie Delta, Canada. Polarizer only. P = secondary porosity; Q = quartz grains and syntaxial quartz overgrowth; S = siderite (very high relief).

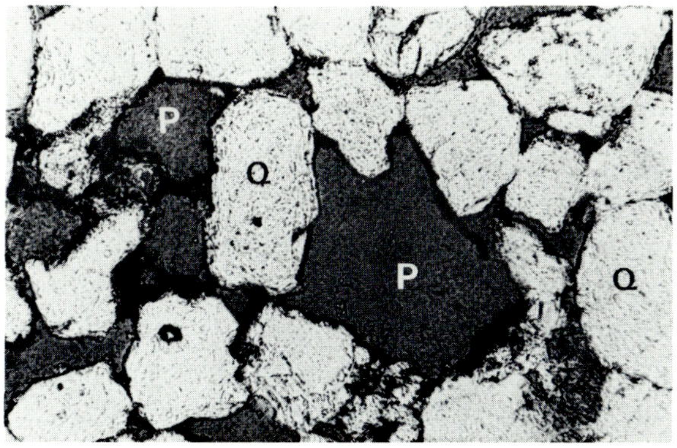

FIG. 70 - Complete dissolution of sedimentary eogenetic and mesogenetic carbonate in quartz arenite, following after complete loss of reducible primary porosity. Mature Zone. Lower Cretaceous, Scotian Shelf, Canada. Polarizer only. P = secondary porosity; Q = quartz.

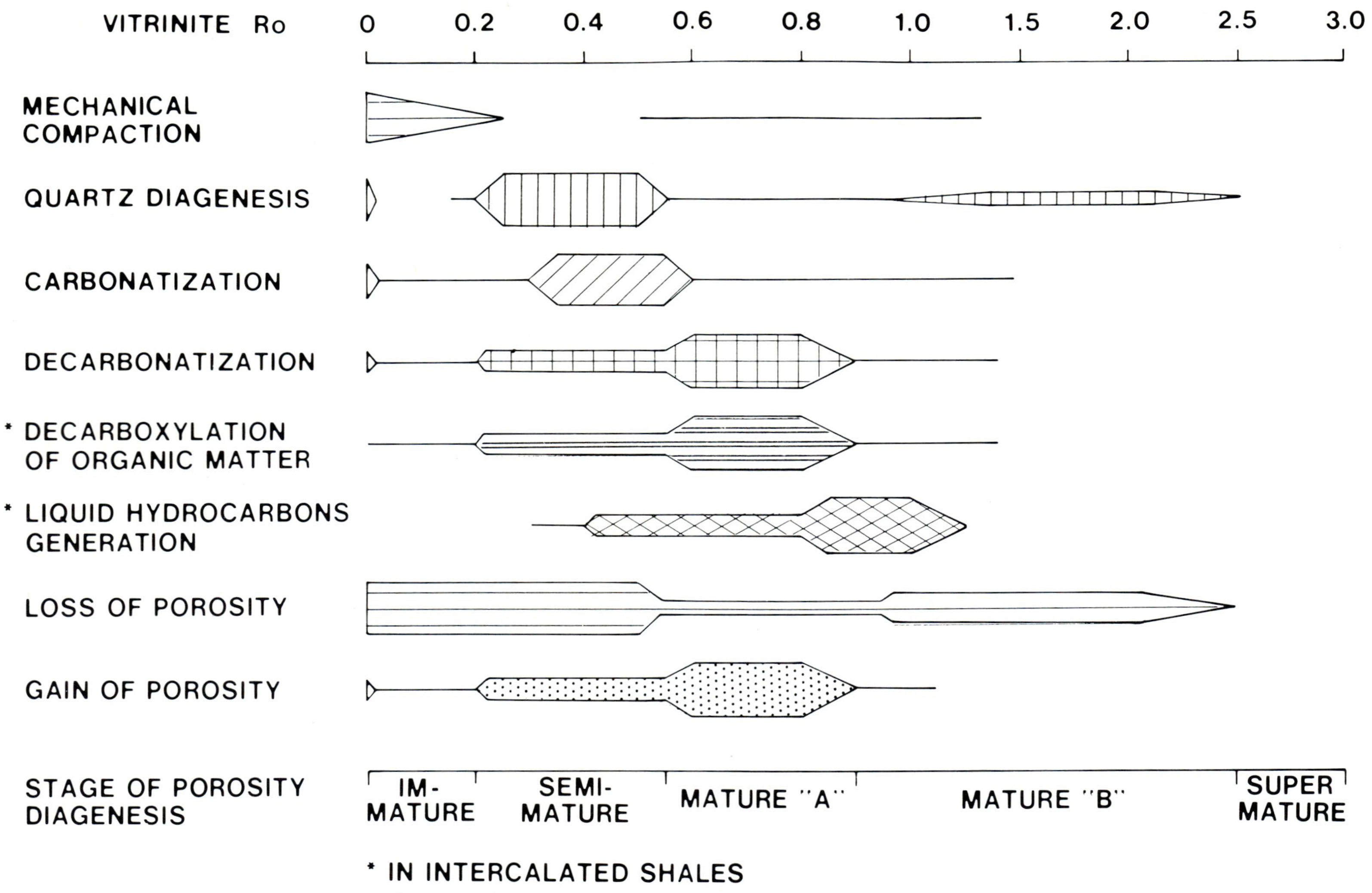

FIG. 71 - Burial diagenesis of quartz arenites.

Vitrinite, if associated with quartz arenites, records their temperature-time exposure by its change in reflectance which can be used as a scale for the thermomaturation of quartz arenites. Thus specific reflectivity values of vitrinite correspond to textural stages of mesodiagenesis.

Mechanical compaction during the immature stage mainly affects primary porosity, reducing it by about one quarter. Mesogenetic carbonatization, decarbonatization, and decarboxylation of organic matter in intercalated shales operate at extremely low levels during the immature stage.

Quartz diagenesis, that is, the chemical compaction of quartz arenites and silica cementation, operates most strongly during the semi-mature stage, mainly reducing primary porosity. Carbonatization reaches its maximum during the last two-thirds of the semi-mature stage, chiefly filling primary porosity but also a considerable volume of secondary porosity. Decarbonatization and decarboxylation are fairly active throughout the semi-mature stage. Moderate generation of liquid hydrocarbons in intercalated shales may take place during the latter part of the semi-mature stage. Loss of porosity predominates over porosity gain though a considerable amount of secondary porosity is formed. All effective primary porosity has been destroyed by the end of the semi-mature stage but an appreciable amount of secondary porosity does exist.

The mature stage "A" is characterized by maximum decarbonatization and decarboxylation of organic matter. Generation of liquid hydrocarbons is, for the most part, moderate but increases to maximum levels towards the end of the mature stage "A". Carbonatization, quartz diagenesis and mechanical compaction operate at low levels. Gain of secondary porosity appears to be at its maximum and results mainly from decarbonatization. It exceeds loss of secondary porosity by a wide margin.

The initial phase of mature stage "B" shows low levels of carbonatization, decarboxylation, decarbonatization, mechanical compaction, and quartz diagenesis, whereas the generation of liquid hydrocarbons continues for a short period, at its maximum level before decreasing. Loss of secondary porosity is minor and gain of secondary porosity is very low.

Quartz diagenesis is the only significant diagenetic activity during the intermediate and final phases of the mature stage "B". It eventually eliminates any reducible secondary porosity. Organic metamorphism generates large volumes of methane during the intermediate phase of the mature stage "B".

The zone of maximum active decarbonatization underlies the zone of maximum active carbonatization. It appears that the former is the source of carbonate for the latter. This means that carbonate must be moving upward in aqueous solution to form the "carbonate curtain." The reasons for the precipitation of carbonate at the zone of carbonatization have not yet been satisfactorily determined. It appears as if the diagenesis of clay minerals in the intercalated shales may influence the chemistry of interstitial waters (change of ionic ratios, alkalinity, pH values) in such a way as to cause carbonate precipitation. Alternatively, formation waters which migrate upwards from a geopressure zone into zones of hydrostatic pressure could precipitate carbonates in the zone of pressure transition.

The occurrence of decarbonatization in most sedimentary basins corresponds very closely with the rate of production of carbon dioxide in intercalated shales or coals resulting from decarboxylation of maturing organic matter. Decarbonatization is most intense in quartz arenites which are closely associated with maturing organic-rich hydrocarbon source rocks. It appears as if generation of carbonic acid in the subsurface is the prime cause of mesogenetic decarbonatization. Other apparently common causes for mesogenetic decarbonatization include: (1) physicochemical gradients at the boundaries of geopressure zones; (2) release of water from clay minerals (Savkevic, 1971); and, (3) generation of hydrogen sulfide.

In many instances mesogenetic decarbonatization takes place in very tight quartz arenites. However, fractures and the irreducible lamellar porosity apparently provide sufficient access for the leaching brines to initiate decarbonatization. As secondary porosity is being created, it continuously improves access for the acid waters. The percentage of secondary porosity which can be generated by complete decarbonatization is limited both by the carbonate content of the sandstone and by the mechanical strength of the non-carbonate framework.

Quartz arenites containing carbonate may escape decarbonatization for one of the following reasons: (1) the subsurface environment does not generate chemical conditions for leaching, e.g. lack of organic matter; (2) the leaching capacity of the migrating formation water has been exhausted previously; and, (3) the leaching formation water has no access to the quartz arenite strata.

In the development of a sedimentary basin intensive decarbonatization will likely occur soon after the oldest quartz arenites have reached the mature zone. These quartz arenites contain mainly sedimentary and eogenetic carbonate. Reprecipitation of this dissolved carbonate in the overlying semi-mature zone may form a thin initial "carbonate curtain." Continued subsidence and sedimentation eventually moves this initial "carbonate curtain" into the mature zone where it may be subjected to strong decarbonatization which, in turn, could lead to a thicker "carbonate curtain" in the quartz arenites which have newly arrived in the semi-mature zone. In this manner both the zone of maximum decarbonatization and maximum carbonatization can move continually into younger sediments. In the process much of the carbonate may be recycled many times over and large volumes of carbonate may accumulate in the moving zones of active carbonatzation, provided that sufficient organic matter is present throughout the sedimentary sequence to sustain decarbonatization.

Most liquid hydrocarbons and probably most of the natural gas is generated in the mature zone of burial diagenesis after the peak of development of secondary porosity. This places quartz arenites with secondary porosity in a very favorable situation to trap these hydrocarbons. Indeed, it is reasonable to assume that primary migration emplaced most oil and gas found in secondary sandstone porosity (Figures 65 & 66), whereas secondary migration supplied much of the hydrocarbons encountered in primary porosity.

THE POROSITY DIAGENESIS OF SANDSTONES OF INTERMEDIATE AND LOW MINERALOGICAL MATURITY

Sandstones of lesser mineralogical maturity than quartz arenites exhibit a similar relationship between the maturation of organic matter and the development of mesogenetic decarbonatization porosity as described above. However, both primary and secondary porosity are reduced by chemical compaction at lower temperatures and at faster rates than observed in quartz arenites. Chemical constituents derived from the dissolution of minerals such as feldspars, micas, amphiboles, and pyroxenes precipitate as mesogenetic cements mainly consisting of clay minerals and zeolites (Hayes, 1979). Chemical compaction and associated intrastratal precipitation of cements may collectively be termed framework diagenesis, analogous to quartz diagenesis. Mesogenetic carbonatization in these sandstones involves more replacement and less cementation when compared with quartz arenites. Carbonatization tends to culminate during the early phase of mature stage "A" at about the same level of thermomaturation as in quartz arenites.

In general, the development and disappearance of mesogenetic secondary porosity follows a similar pattern in sandstones of intermediate and low mineralogical maturity to that observed in quartz arenites. The detail of the diagenetic histories of sandstones of lesser mineralogical stability is often very complex and much more work remains to be done to establish the basic relationships as clearly as is now possible with quartz arenites. The Cretaceous sandstones of the Labrador Shelf, the Tertiary sandstones of the Mackenzie Delta, and the Mesozoic and Tertiary sandstones of the Cook Inlet are prime candidates for such work. These sandstones of intermediate to low mineralogical maturity contain large hydrocarbon accumulations in reservoirs of secondary sandstone porosity.

EXAMPLES OF POROSITY DISTRIBUTION

Plots of sandstone porosity versus depth are easily constructed using core analyses and/or log interpretations from wells. Petrographic analysis of the sandstones and application of the criteria and concepts developed in this paper allow a much more complete interpretation of such porosity/depth plots than has hitherto been possible.

<u>Mackenzie Delta example</u>. - Figure 72 represents a porosity/depth plot illustrating the rate of decline of primary and secondary sandstone porosity of a well in the Mackenzie Delta which penetrated Tertiary sandstones of intermediate mineralogical maturity and Lower Cretaceous quartz arenites. The porosity of the Tertiary sandstones above a depth of 2,100 meters is predominantly primary and declines rapidly with depth. Primary porosity has been reduced by both framework diagenesis and by carbonate cementation. Both processes are apparently active at the present time. Below 2100 meters the sandstone porosity is essentially secondary and declines with depth at a much slower rate compared with that of primary porosity. Active maximum decarbonatization occurs between 2,400 and 2,500 meters. The relatively high porosities below 4,000 meters do not necessarily indicate decarbonatization at that depth; rather they represent a zone of exceedingly high decarbonatization with porosities probably exceeding 30 volume percent which has subsequently been reduced to the present levels by quartz diagenesis.

Another well in the Mackenzie Delta with similar stratigraphy to the above well shows a wide range of secondary porosity within the narrow depth interval 2,650 to 3,050 meters (Figure 73). This zone is in the mature stage "A" of burial diagenesis and the sandstones from 2,650 to 2,900 meters are gas bearing. The secondary porosity is the result of decarbonatization involving the leaching of sedimentary, eogenetic, and mesogenetic carbonate (Figures 74, 75, 76,). High original carbonate content and nearly complete decarbonatization are the attributes of the sandstone intervals with high secondary porosity. Low secondary porosities are shown by sandstone intervals which either originally contained only moderate amounts of carbonate or experienced incomplete decarbonatization (Figures 77, 78, 79). Accumulation of the gas occurred

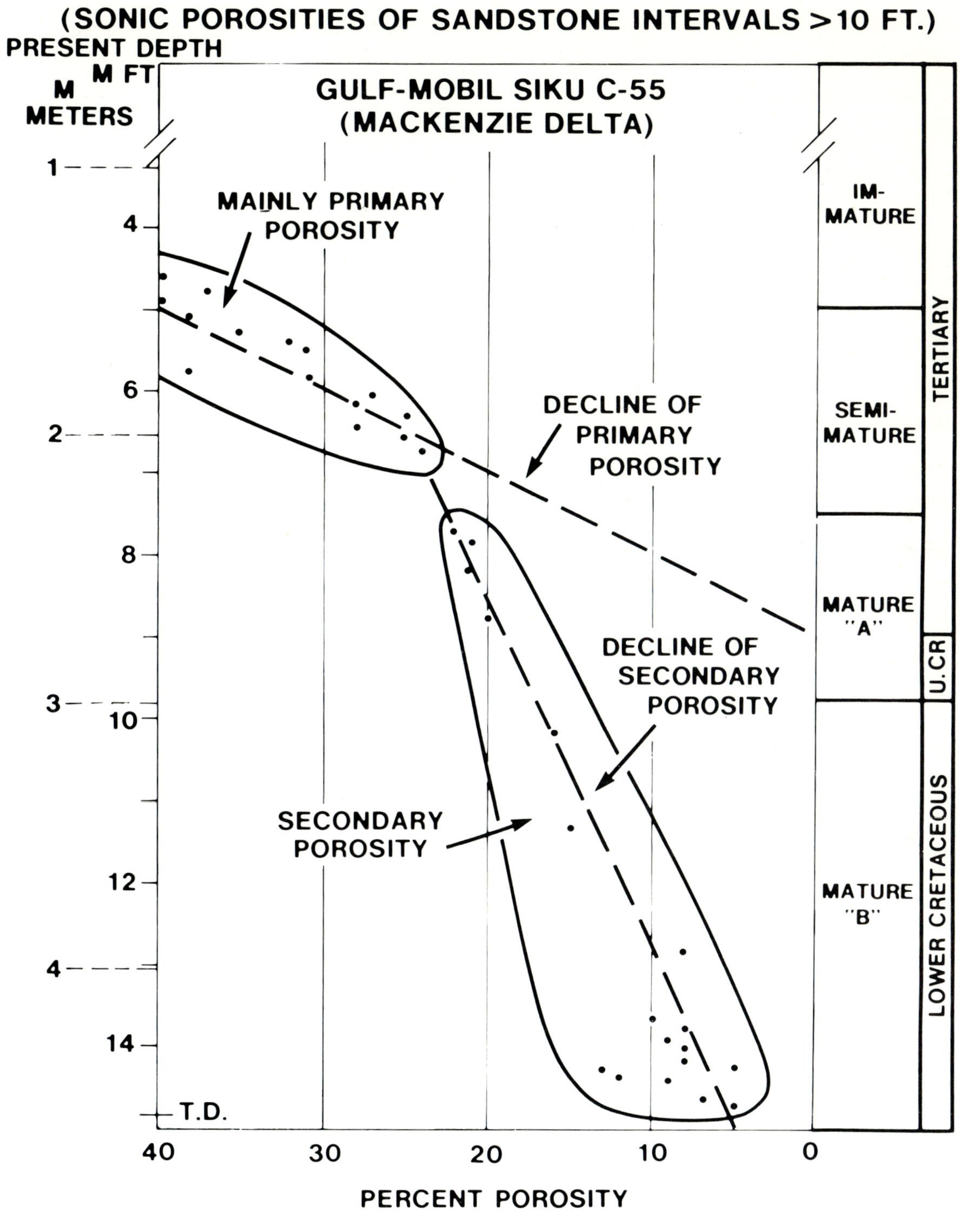

FIG. 72 - Mackenzie Delta: Depth/porosity plog showing decline of primary and secondary porosity (sonic porosities of sandstone intervals > 10 ft.).

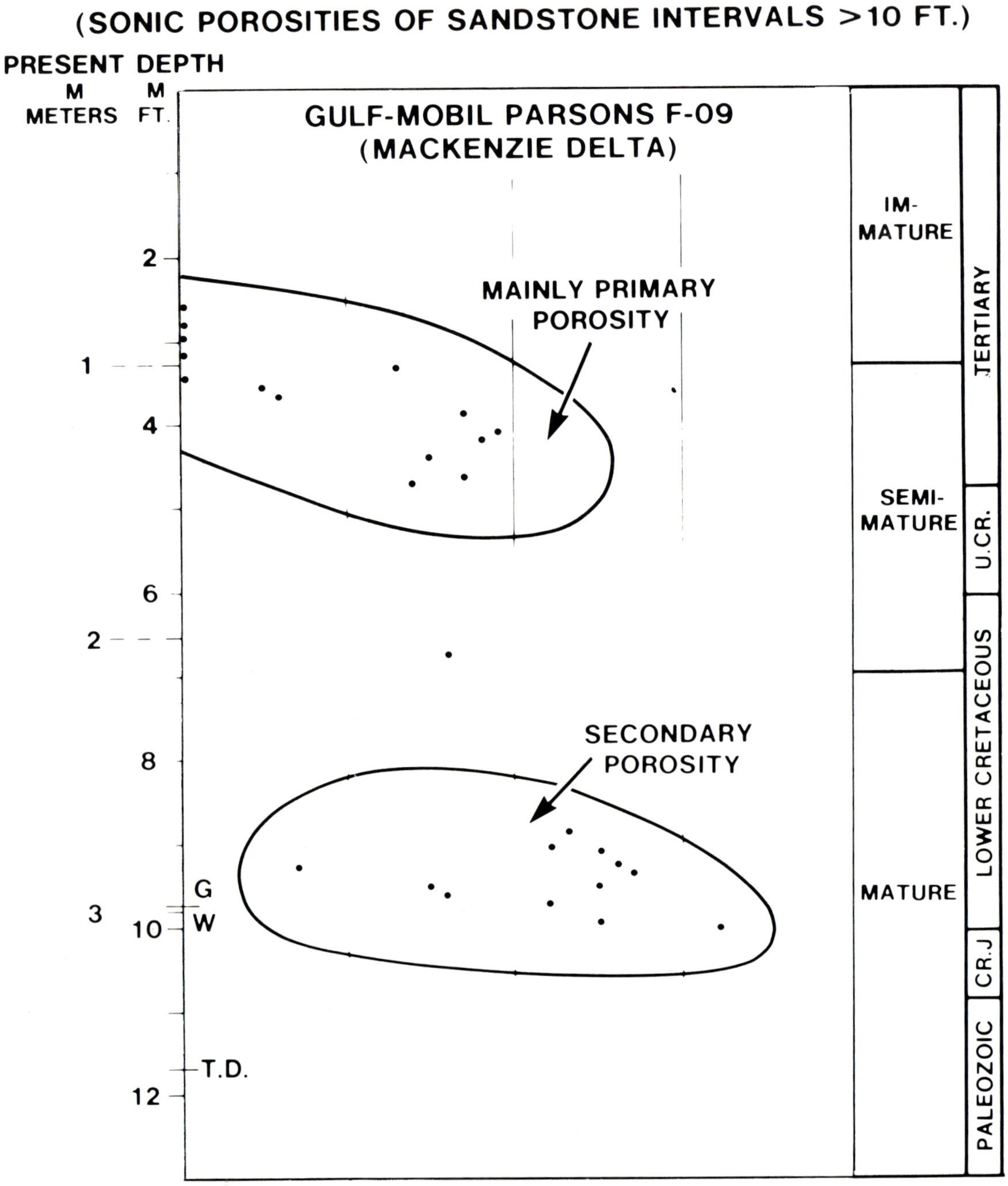

FIG. 73 - Mackenzie Delta: Depth/porosity plot showing wide range of secondary porosity (sonic porosities of sandstone intervals >10 ft.).

FIGS. 74 to 76 - Porosity variation in a sandstone zone of mature textural mesodiagenesis, Lower Cretaceous, Mackenzie Delta.

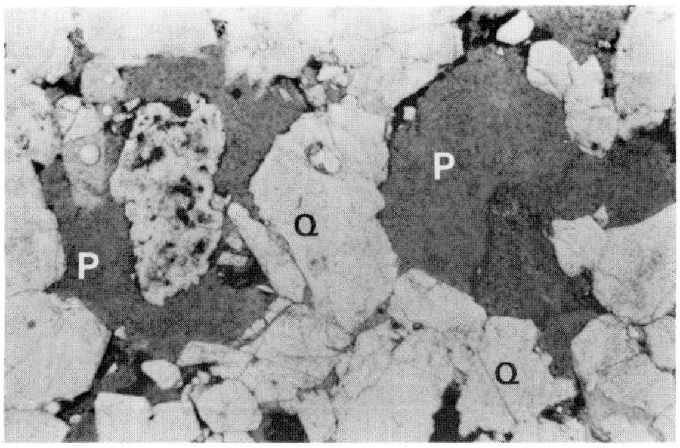

FIG. 74 - Secondary porosity created by mesogenetic leaching of mainly sedimentary and eogenetic carbonates. Mature stage. Lower Cretaceous quartz arenite, Mackenzie Delta, Canada. Polarizer only. P = secondary porosity; Q = quartz.

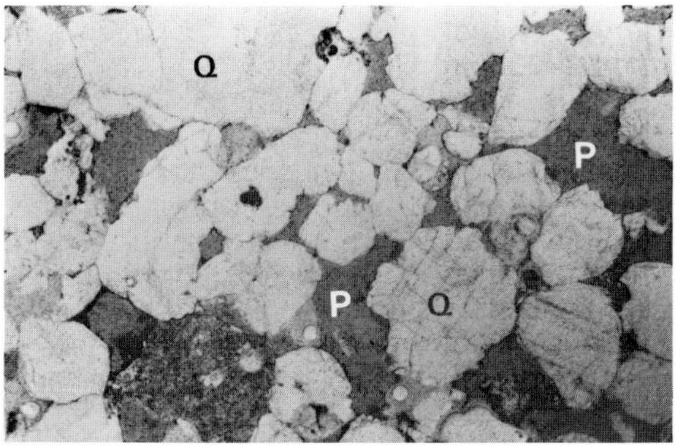

FIG. 75 - Secondary porosity created by mesogenetic leaching of sedimentary porosity, eogenetic carbonate and mesogenetic intergranular carbonate cement. Mature stage. Lower Cretaceous quartz arenite, Mackenzie Delta, Canada. Polarizer only. P = secondary porosity; Q = quartz.

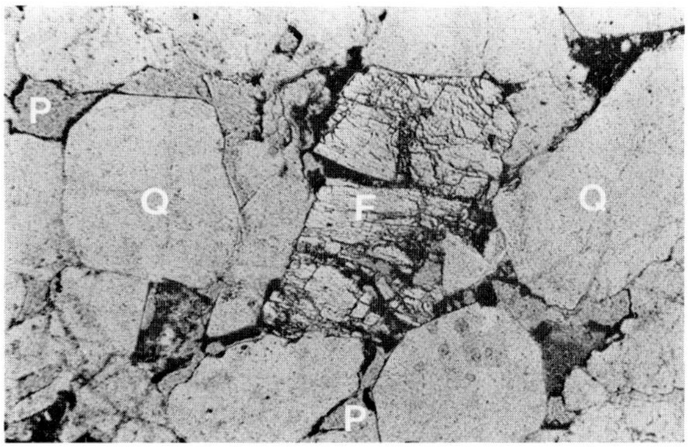

FIG. 76 - Secondary porosity created by mesogenetic leaching of mainly mesogenetic intergranular carbonate cement. Feldspar grains were fractured after leaching of mesogenetic carbonate that partially replaced the feldspar. Mature stage. Lower Cretaceous, subfeldspathic arenite, Mackenzie Delta, Canada. Polarizer only. P = secondary porosity; Q = quartz; F = feldspar (fractured grains).

FIGS. 77 to 79 - Porosity variation in a sandstone zone of mature textural mesodiagenesis Lower Cretaceous, Mackenzie Delta

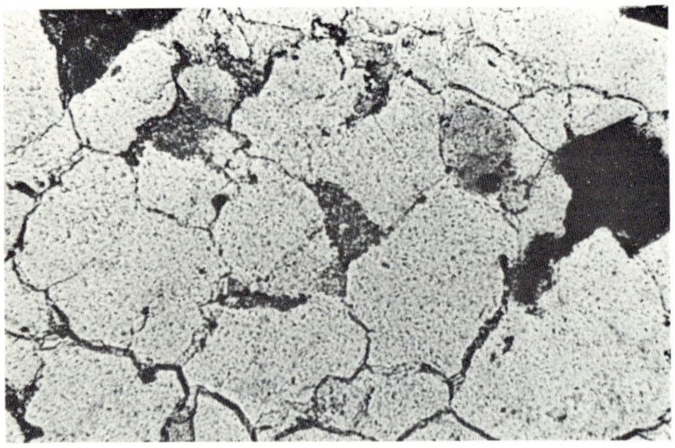

FIG. 77 - Lack of effective porosity due to absence of carbonate when sandstones entered the mature stage. Lower Cretaceous quartz arenite. Mackenzie Delta, Canada. Polarizer only.

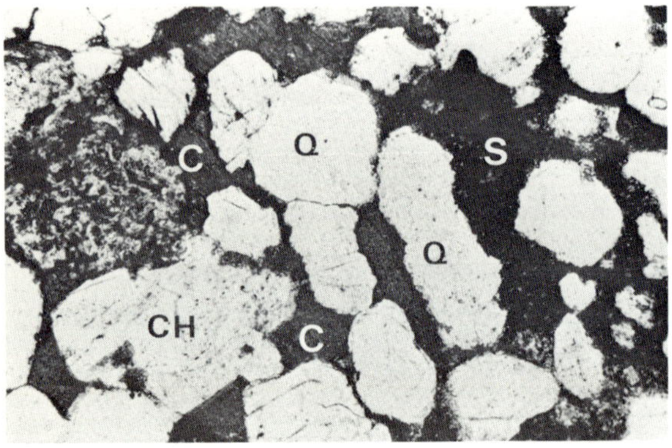

FIG. 78 - Lack of effective porosity due to absence of mesogenetic leaching of sedimentary eogenetic and mesogenetic carbonate. Mature stage. Lower Cretaceous sublithic arenite. Mackenzie Delta, Canada. Polarizer only. Q = quartz; S = siderite; CH = chert; C = mesogenetic calcite.

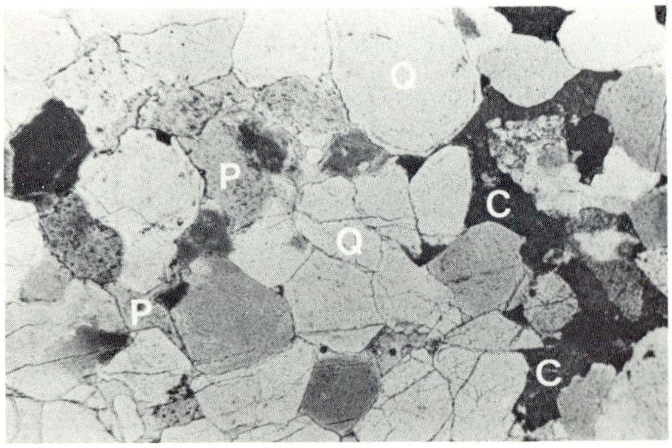

FIG. 79 - Low effective porosity due to incomplete mesogenetic leaching of mesogenetic calcite. Mature stage. Lower Cretaceous quartz arenite. Mackenzie Delta, Canada. Half crossed nicols. P = secondary porosity; Q = quartz grains and mesogenetic syntaxial quartz overgrowth; C = mesogenetic poikilotopic calcite cement.

after decarbonatization during the mature stage "A" of burial diagenesis. The presence of hydrocarbons had little or no effect on the preservation of porosity. However, if the gas bearing sandstones were to be buried deeply enough to enter mature stage "B" of burial diagenesis, then the presence of hydrocarbons would prevent quartz diagenesis and preserve porosity.

Scotian Shelf example. - Porosities of Cretaceous and Upper Jurassic sandstones of a well on the Scotian Shelf are plotted versus depth in Figure 80. The sandstones are principally composed of quartz arenites, feldspathic quartz arenites, and glauconitic quartz arenites, and originally contained considerable amounts of sedimentary and eogenetic carbonate. Intercalated shales vary greatly in their mineralogy and contain very low to moderately high amounts of organic matter of widely varying composition.

Primary and secondary porosity overlap throughout the semi-mature stage of burial diagenesis to form a gradation from very high primary porosities at the top to high secondary porosities at the bottom. Tight sandstones containing abundant mesogenetic carbonate cement are intercalated with the porous sandstones which experienced considerable decarbonatization (Figures 81, 82, 83). Thus multiple zones of carbonatization alternate with multiple zones of decarbonatization possibly because of the variable content and type of organic matter and lithology of the intercalated shales. It is likely that a number of horizons of carbonatization and decarbonatization are presently active.

In mature stage "A" of burial diagenesis all sandstones containing carbonate experienced renewed decarbonatization to some degree, whereas carbonatization was negligible (Figures

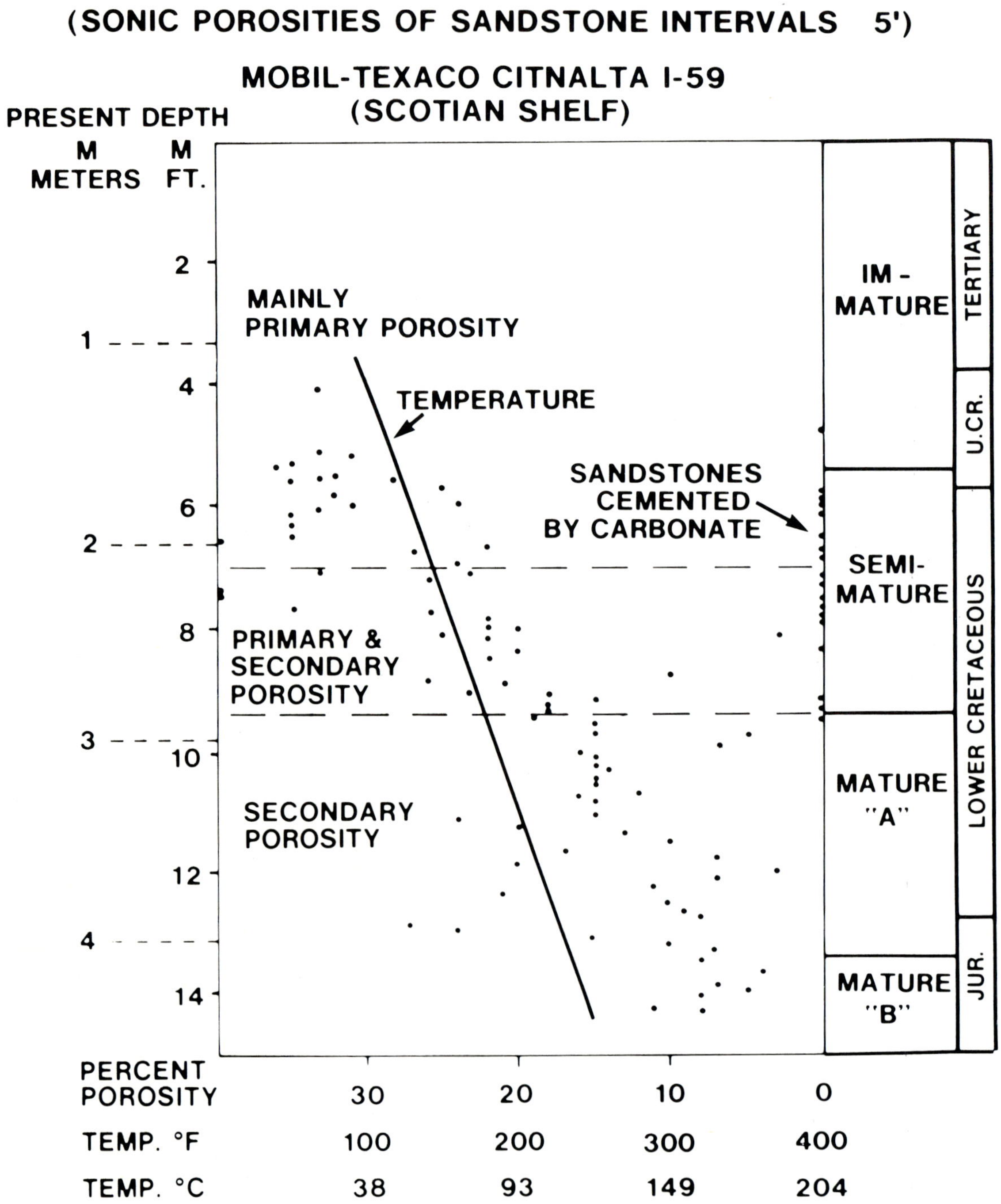

FIG. 80 - Scotian Shelf: Depth/porosity plot showing overlap of primary and secondary porosity (sonic porosities of sandstone intervals >10 ft.).

FIGS. 81 to 83 - Sequence of stages of textural mesodiagenesis with depth in Cretaceous sandstones of the Scotian Shelf

FIG. 81 - Immature stage. High primary porosity in Cretaceous quartz arenite. Mobil Citnalta I-59, Scotian Shelf, Canada. Polarizer only. P = primary porosity; Q = quartz grains.

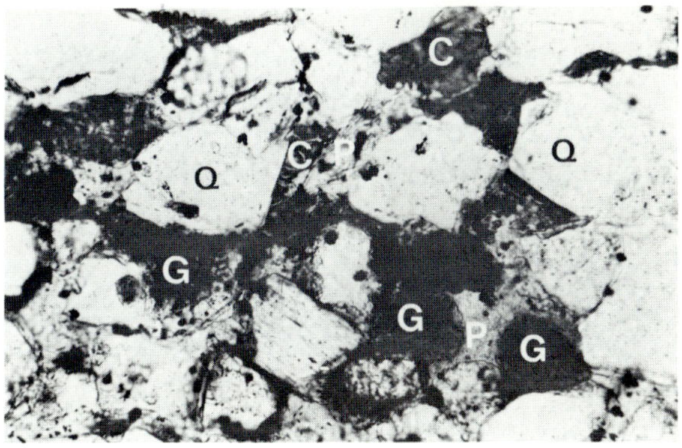

FIG. 82 - Semi-mature stage. Primary and secondary porosity in glauconite quartz arenite. Incomplete mesogenetic cementation by poikilotopic calcite was followed by partial mesogenetic leaching of calcite. Cretaceous. Mobil Citnalta I-59, Scotian Shelf, Canada. Polarizer only. P = primary and secondary porosity; Q = quartz grains and mesogenetic syntaxial quartz overgrowth; C = calcite; G = sedimentary and eogenetic glauconite.

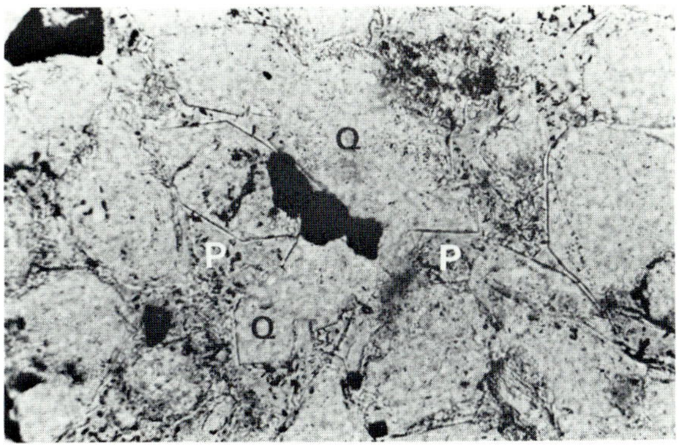

FIG. 83 - Semi-mature stage. Cementation of secondary and primary porosity by mesogenetic syntaxial quartz overgrowth. Cretaceous quartz arenite. Mobil Citnalta I-59, Scotian Shelf, Canada. Polarizer only. P = primary and secondary porosity; Q = quartz.

FIGS. 84 to 86 - Sequence of stages of textural mesodiagenesis with depth in Cretaceous sandstones of the Scotian Shelf

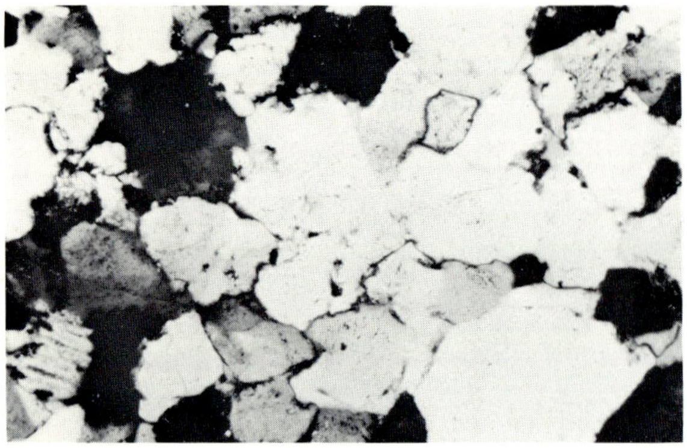

FIG. 84 - Mature stage. Lack of effective porosity due to absence of carbonate when sandstone entered the mature stage. Cretaceous quartz arenite. Mobil Citnalta I-59, Scotian Shelf, Canada. Crossed nicols.

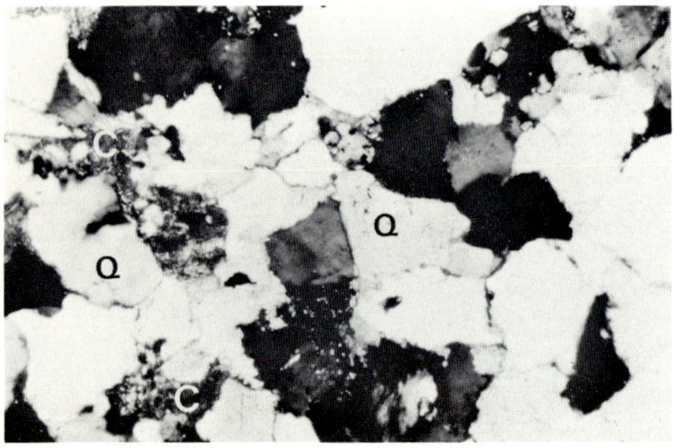

FIG. 85 - Mature stage. Lack of effective porosity due to absence of mesogenetic leaching of mesogenetic calcite cement. Cretaceous quartz arenite, Mobil Citnalta I-59, Scotian Shelf, Canada. Crossed nicols. Q = quartz grains and mesogenetic syntaxial quartz overgrowth cement; C = mesogenetic poikilotopic calcite cement.

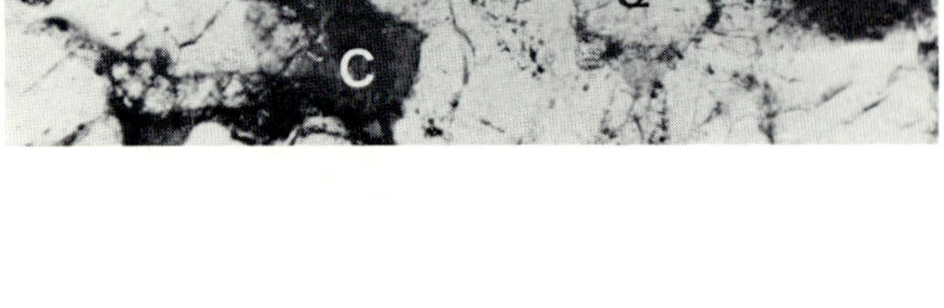

FIG. 86 - Mature stage. Presence of secondary porosity due to mesogenetic leaching of mesogenetic poikilotopic calcite. Cretaceous subfeldspathic arenite. Mobil Citnalta I-59, Scotian Shelf, Canada. Polarizer only. P = secondary porosity; Q = quartz grains and mesogenetic syntaxial quartz overgrowth cement; C = mesogenetic calcite cement and calcite replacing detrital grains and quartz cement along the boundaries.

84, 85, 86). The sandstones with high secondary porosities are often closely associated with hydrocarbon source rocks, and contain condensate and natural gas. Quartz diagenesis and framework diagenesis of other grains is responsible for the decline of secondary porosity in the mature stage "B" of burial diagenesis.

RESERVOIR ASPECTS

Pore geometry. - The geometry of secondary pores if compared to the geometry of primary precursor porosity can be described in three basic morphologies.

(1) Geometry identical to precursor primary porosity.
(2) Geometry modified from precursor primary porosity.
(3) Geometry unrelated to precursor primary porosity.

Identical geometry is shown by exact replicas of unaltered primary pores or of residual primary pores. (Figure 87). The latter may have been reduced by such processes as compaction, pressure solution and partial cementation by silica overgrowth. These secondary pores occur wherever the soluble precursor consisted only of pore-filling cement. A considerable yet, overall, subordinate amount of secondary porosity falls into this morphologic class. Pore throats of the reopened unaltered primary intergranular porosity tend to be of concave-triangular cross section. In sandstones with reopened residual primary porosity two types of pore connections occur together: pore throats of straight-triangular or concave-triangular cross secion and irreducible pore throats at grain contacts with lamellar cross section.

Modified geometry occurs most commonly as enlarged modified replicas of the precursor primary porosity and

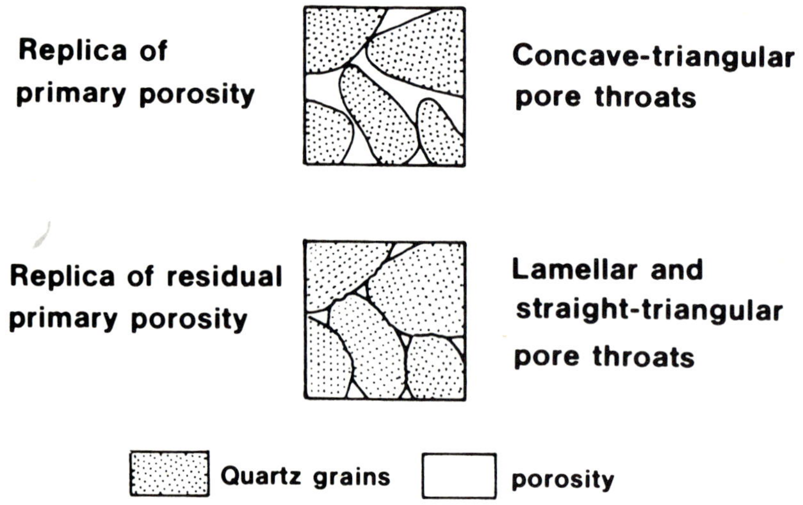

FIG. 87 - Geometry of secondary porosity identical to precursor primary porosity.

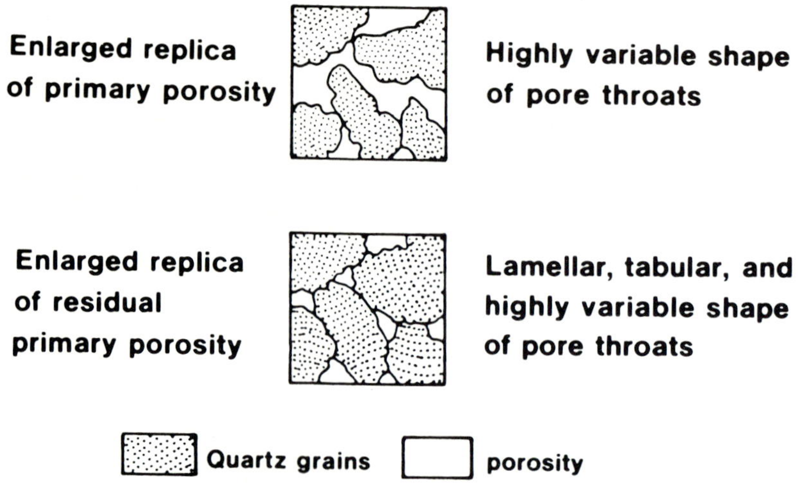

FIG. 88 - Geometry of secondary porosity modified from precursor primary porosity residual primary porosity (Figure 88). In these cases not only soluble intergranular cement was removed but also margins of grains and grain overgrowths. For example, dissolution may have affected directly the margins of grains or the soluble minerals that marginally replaced the grain framework. Modified pore geometry shows pore throats of highly variable shape if the precursor was normal primary porosity. In the case of enlargement of a residual primary porosity fabric, the spectrum of pore throats is also quite varied and includes lamellar, tabular, tubular prismatic and irregular shapes in three dimensions. Prismatic pore throats of modified geometry are commonly convex-triangular in cross section.

Unrelated geometry is shown by a considerable percentage of secondary pores. Six types of pores are shown in Figure 89 that are unrelated in their fabric to pre-existing primary pores:

(1) Molds of soluble minerals which replaced quartz and silicate grains along grain-to-grain contacts are common in sandstones which first lost their primary porosity and then experienced marginal replacement of grains along the irreducible lamellar porosity between the grains. Pores and pore throats are intersecting-tabular in shape forming often a well connected network of porosity.

(2) Molds of grains and noncementing crystals reflect the geometry of their precursor and show predominantly convex-equant cross sections. Molds provide pore throats only if they are so abundant that they adjoin each other. In this case they may form a continuous network of interconnecting channels with pore throats of highly irregular shape. In most instances, however, the shape of pore throats is not influenced by the presence of grain and crystal molds and depends entirely on the

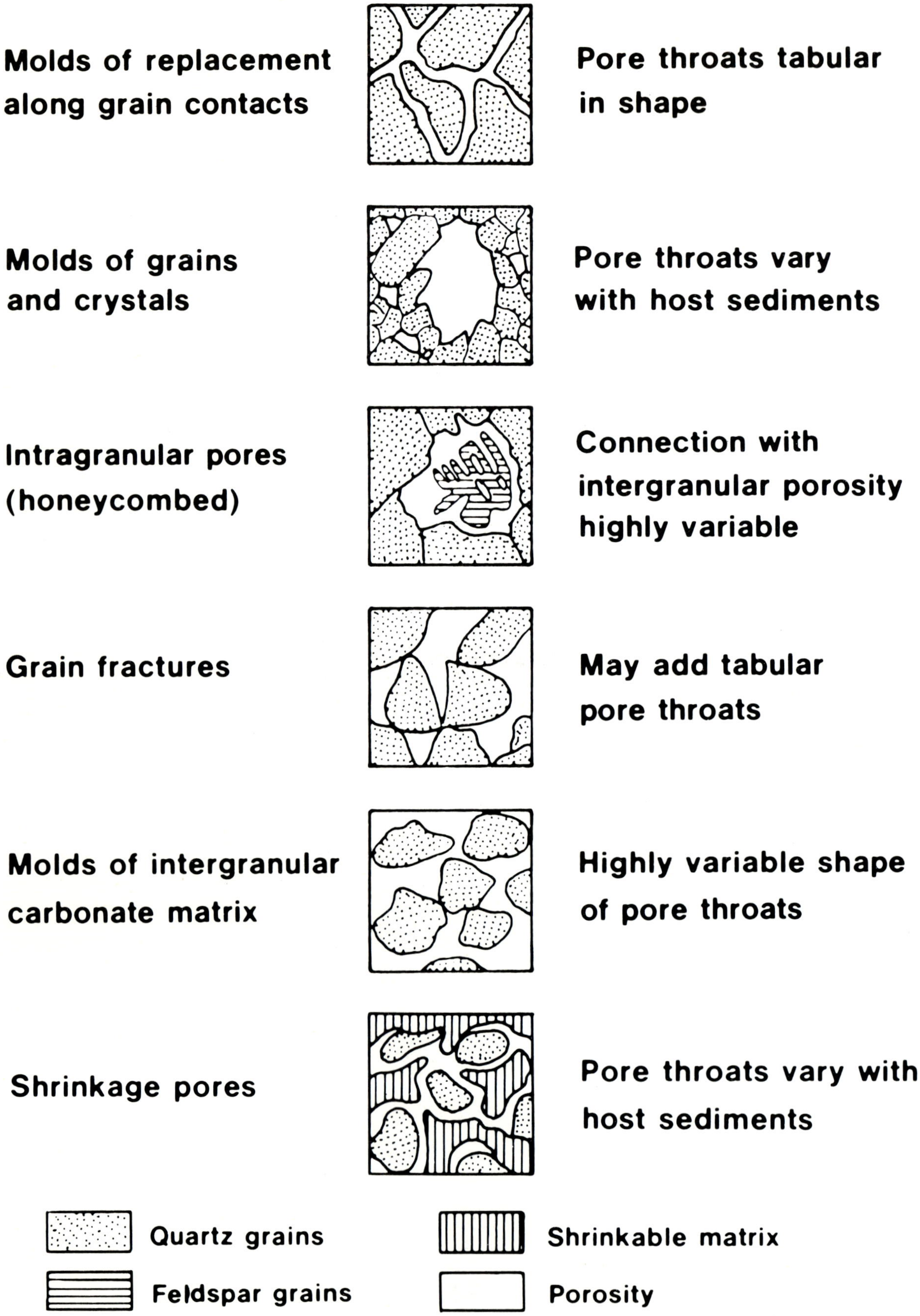

FIG. 89 - Geometry of secondary porosity unrelated to precursor primary porosity pore geometry of the sandstone host encasing the molds.

(3) Intragranular pores are often unrelated to primary porosity. They appear in a vast spectrum of geometric shapes and may or may not have effective interconnection with pore systems beyond the grain boundary. Rarely do intragranular pores function as important throats of an effective network of porosity.

(4) Grain fractures have lamellar, tabular or wedge-shape morphology and may provide effective pore throats with tabular and cuneiform cross section.

(5) Molds of intergranular carbonate matrix appear most commonly in the same morphology as that of enlarged replicas of primary porosity. This is due to the fact that chemical compaction or replacement occurs at the interface between matrix and grain framework. Patches of sandstone with an original fabric of matrix support are the precursor of oversized pores. These may have a geometry similar to grain molds.

(6) Shrinkage voids form diverse pore morphologies and include partial intergranular, intramatrix and intragranular pores. Characteristic are sinuous, crescent-shaped and lenticular geometric forms. Pore throats appear commonly lenticular and arcuate in cross section.

In summary, the geometry of secondary porosity differs from that of primary sandstone porosity, as follows:

(1) The size range of secondary pores is wider and pore shapes are more diverse.
(2) The spectra of sizes and shapes of secondary pore

throats are radically different.

(3) The porosity distribution is much less uniform in most sandstones with secondary porosity.

Reservoir geometry. - The geometry of sand bodies is usually only one of several factors that determine the configuration of bodies of effective secondary porosity in sandstones showing mature textural diagenesis with complete loss of primary porosity. Only rarely does effective secondary porosity occur throughout the total rock volume of a sand body. Economic hydrocarbon accumulations may occur in sand bodies in which less than 5 percent of the rock volume have effective secondary porosity. Factors which affect the geometry of reservoirs with secondary porosity include: (1) distribution of the soluble minerals that were precursors of secondary porosity; (2) distribution of certain rock fabrics that either impeded or facilitated the development of secondary porosity; (3) position and frequency of conduits for dissolving fluids, e.g. faults and master joints; (4) chemical activity and amount of dissolving fluids that reacted with the sand body; and, (5) distribution of any zones within the sand body where secondary porosity was destroyed. Effective secondary porosity may be continuous within a particular sandstone body or it may occur in several isolated pockets. Both lateral and vertical changes in secondary porosity may be extreme, e.g. a thick continuous sandstone unit may show no development of secondary porosity and be tight in a location that is very close to an excellent reservoir in the same sandstone unit. The geometry of reservoirs with secondary porosity must receive special consideration wherever such reservoirs are exploration targets, are undergoing development drilling, are subject to a stimulation program or are scrutinized by reservoir engineers designing an exploitation program.

Petrophysical characteristics:

As a result of their pore geometry, most sandstones with secondary porosity show strikingly different petrophysical characteristics if compared with sandstones of similar lithology but with primary porosity.

Permeability in sandstones with secondary porosity may be either higher or lower than that in sandstones with the same amount of primary porosity, depending on the nature of the secondary porosity (Figure 90). Permeability is higher where secondary pore throats are larger and have a shape more suitable for improved pore connection. On the other hand, permeability is lower if the secondary porosity consists mainly of isolated pores such as molds of grains and of patches of precursor matrix.

Sonic travel time is lower in sandstones with secondary porosity because the contacts between grains are better welded. The Cretaceous and Tertiary sandstones of the Mackenzie Delta both show this relationship in exactly the same fashion despite the much higher quartz content of the Cretaceous sandstone (Figure 91).

The irreducible water saturation in gas-bearing sandstones with secondary porosity is often much lower when compared with gas-reservoir sandstones with primary porosity (Figure 92). Sandstones with secondary porosity show irreducible water saturations very similar to those of carbonate rocks with vuggy porosity.

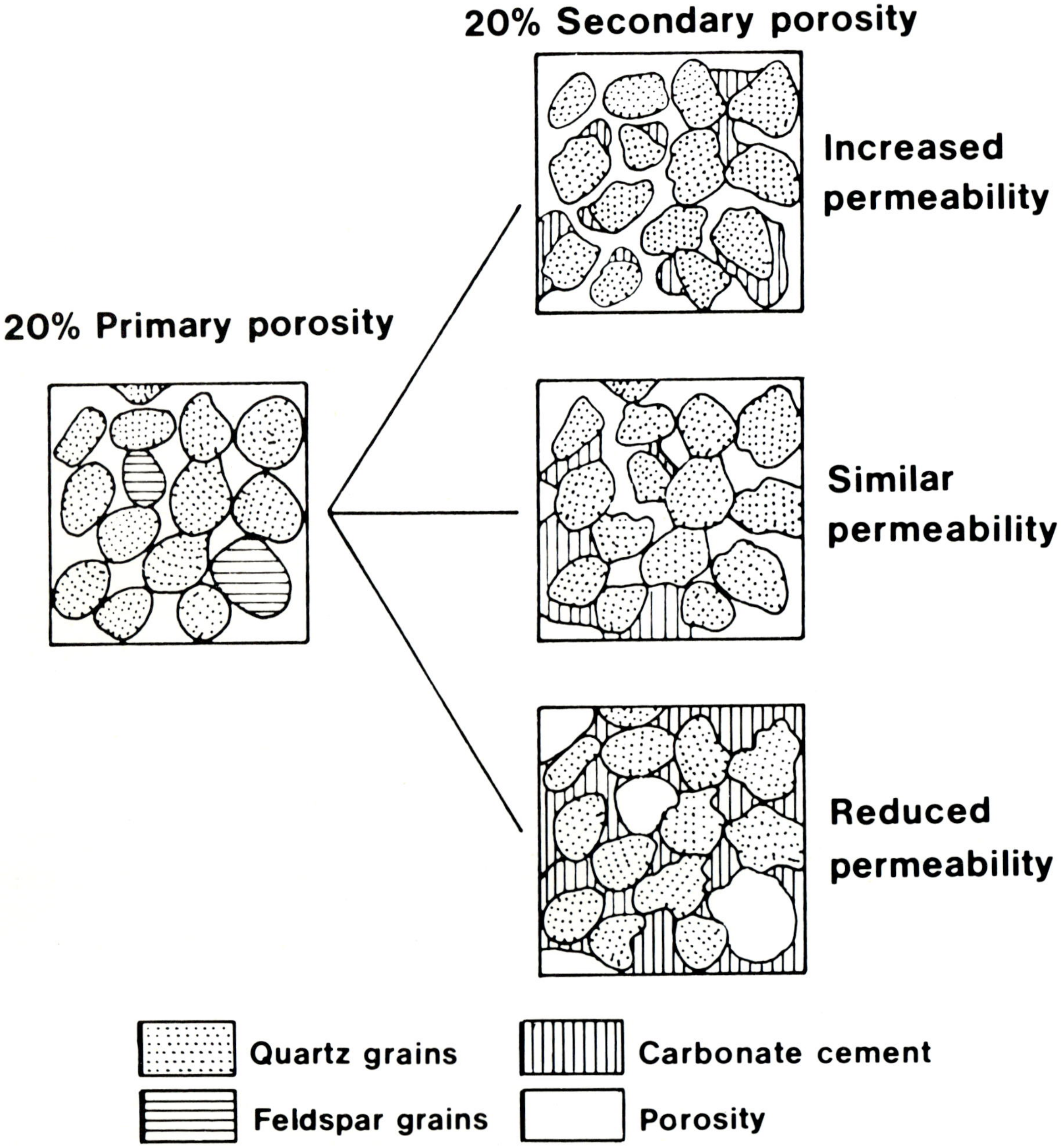

FIG. 90 - Permeability in sandstones with secondary porosity compared with primary porosity.

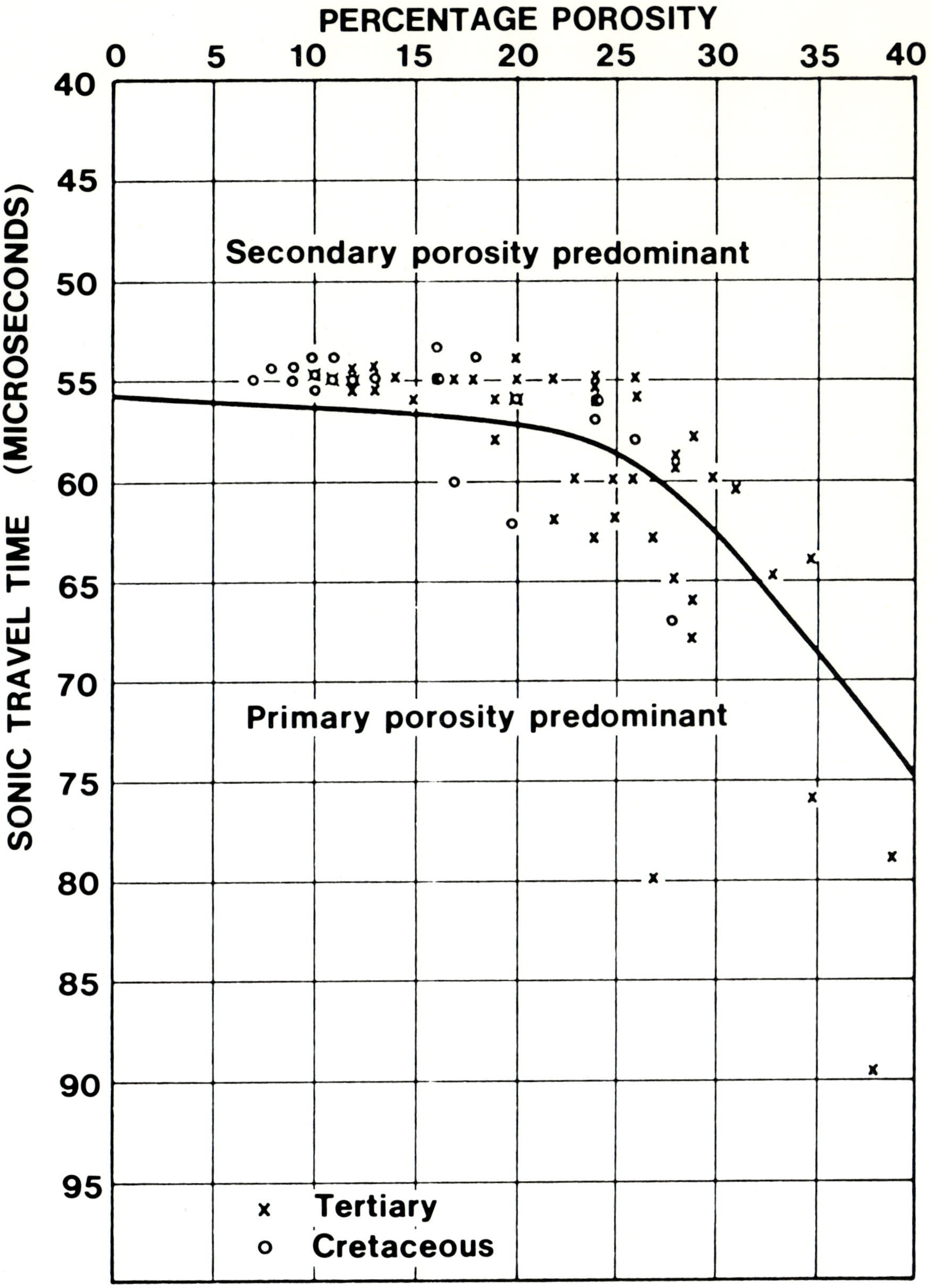

FIG. 91 - Sonic travel time versus porosity from wells in the Mackenzie Deta

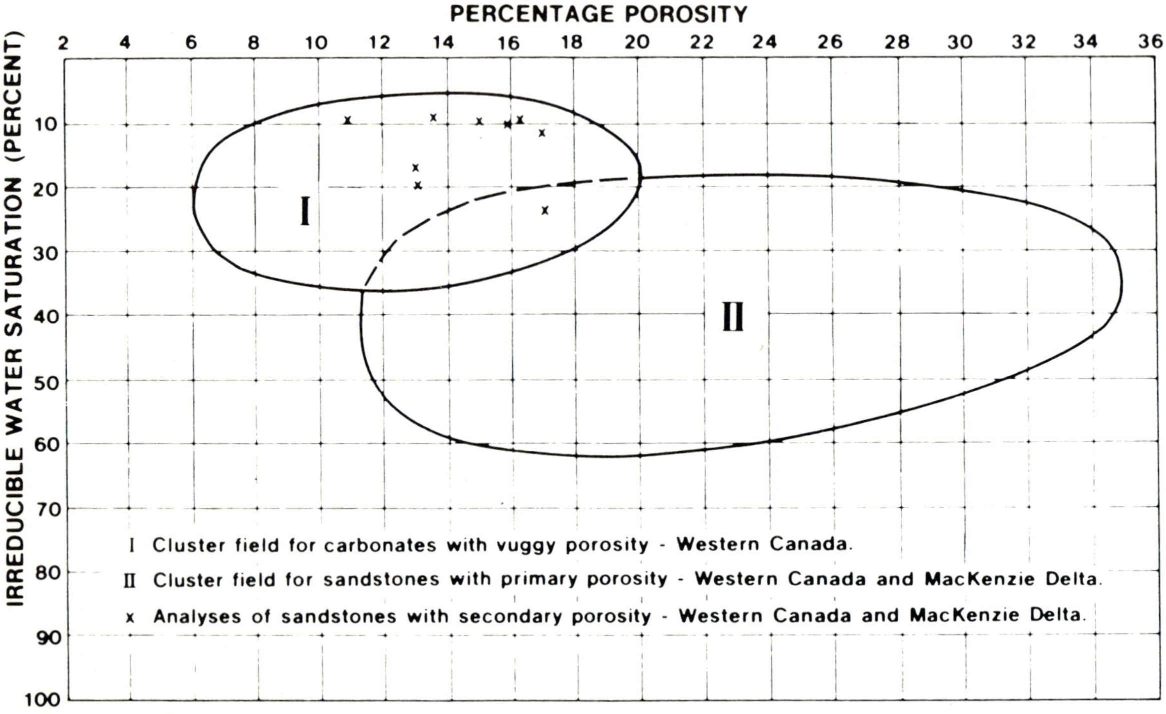

FIG. 92 - Irreducible water saturation versus porosity in gas reservoirs in sandstones with secondary porosity.

Sandstones with secondary porosity generally have a much greater mechanical competence than sandstones with the same volume of primary porosity. Therefore, compaction by overburden pressure will reduce high secondary porosity to a much lesser degree. Sand production should also be much less of a problem in high-porosity sandstones where the porosity is secondary.

Downhole logging methods must, in many cases, respond differently to secondary porosity in sandstones. The sonic log is an established example; other logging methods await evaluation on this point.

<u>General reservoir properties</u>: - The pronounced spatial variability of secondary porosity and permeability in sand-

stones is akin to the variability of porosity and permeability in carbonate rocks. The size of hydrocarbon accumulations in secondary sandstone porosity ranges from subcommercial pools to the largest oil field in North America, the Prudhoe Bay Field. Deliverability of individual well completions range from trivial amounts to several thousand cubic meters per day as shown by the Jurassic fields of the North Sea.

The porosity cut-off for effective porosity varies over a wide range (2 to 25 volume percent). This range is nearly as wide as that for the cut-off for effective porosity in carbonate rocks (2 to 30 volume percent, chalks excepted). The cut-off for effective secondary porosity is in many instances much lower compared to that for effective primary porosity in a sandstone of similar original lithology given comparable conditions of pressure and temperature in the reservoir.

Cementation by clay minerals can be as much a problem in sandstone reservoirs with secondary porosity as it is in reservoirs with primary sandstone porosity. However, response to remedial stimulation techniques can greatly differ from clay-cemented sandstones with primary porosity to those with secondary porosity.

CONCLUSIONS

The following conclusions concerning the role of secondary porosity in the course of sandstone diagenesis can be drawn from the above discussion.

1) Secondary pores in sandstones can be objectively described on the basis of their textural and genetic aspects, and a very large number of porosity types can be defined. Many of the textures of secondary porosity strongly resemble or

mimic those of primary porosity, whereas other textures of secondary porosity differ greatly. In most instances simple petrographic criteria permit the recognition of secondary porosity in thin section. However, the investigation of texture, origin, and alteration of secondary porosity sometimes requires more advanced analytical methods. The quantitative differentiation between coexisting primary and secondary porosity and the establishment of the sequential order of diagenetic events are the most difficult steps in the analysis of secondary sandstone porosity.

2) Several processes are capable of creating significant volumes of secondary sandstone porosity and these include dissolution of a number of minerals, shrinkage of certain minerals, and fracturing. Dissolution of carbonate minerals creates by far the largest volume of secondary sandstone porosity in all three diagenetic regimes.

3) Secondary porosity plays a pre-eminent role in the subsurface diagenesis (mesodiagenesis) of sandstones. In addition, a significant amount of secondary porosity may occasionally develop during early diagenesis (eodiagenesis), and frequently develops during erosional diagenesis (telodiagenesis). The origin and destruction of secondary porosity and associated diagenetic processes are key elements in the diagenesis cycle of sandstone (Figure 93).

4) Decarbonatization during eodiagenesis and telodiagenesis is caused mainly by carbonic acid which is derived from meteoric and biogenic carbon dioxide. Mesogenetic decarbonatization appears to result mainly from leaching by carbonic acid which is derived from the decarboxylation of maturing organic matter in intercalated shales. This decarboxylation occurs over a wide range of temperature-time exposures but most of the

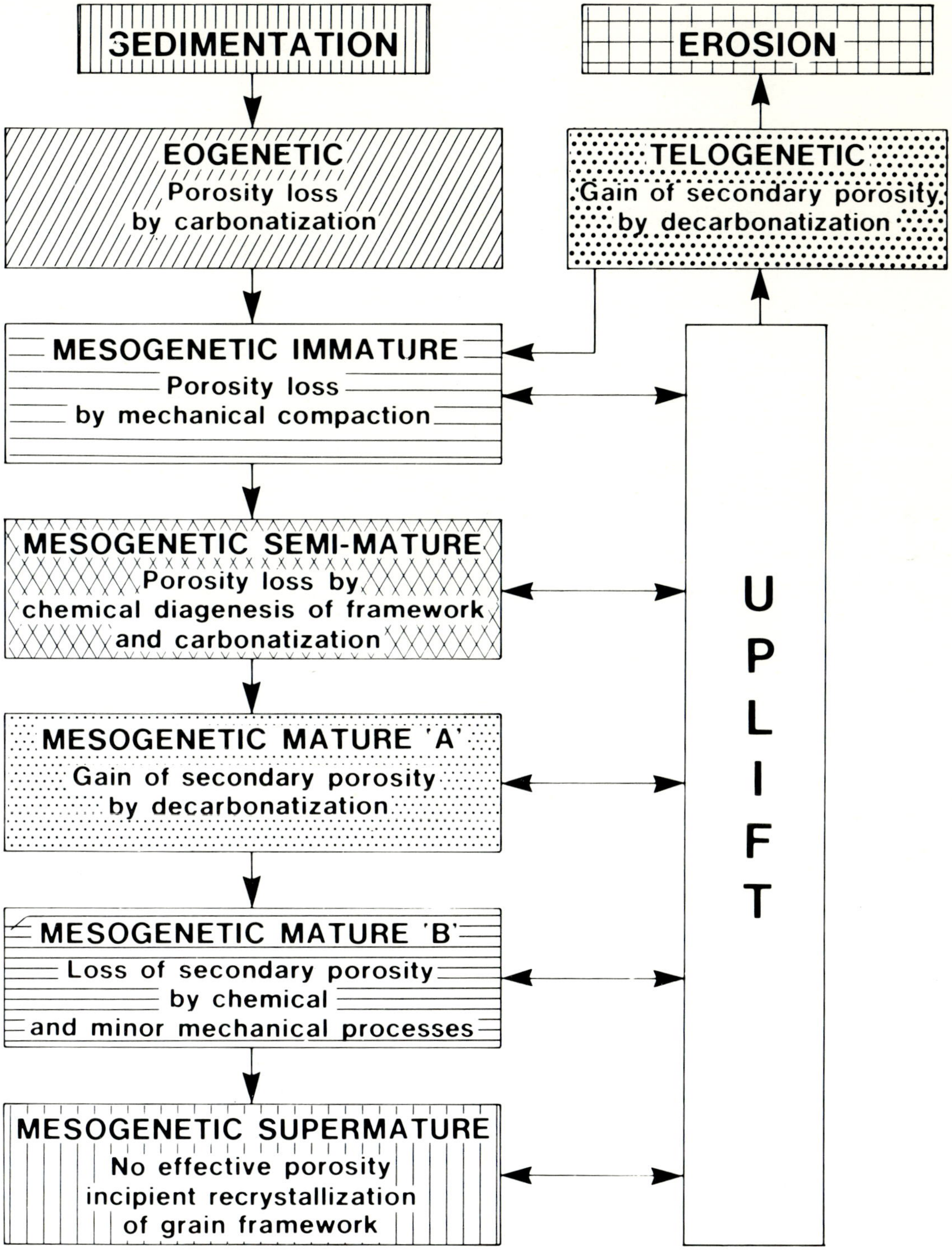

FIG. 93 - Diagenesis cycle of sandstones.

decarboxylation occurs after the sandstones have lost their primary porosity.

5)) Decarbonatization is the source of dissolved chemical constituents which, for the most part, migrate upwards into overlying sandstones and are precipitated there as mesogenetic cement and replacement. In a subsiding prism of accumulating sediments repeated recycling of carbonate may occur leading to very high concentrations of carbonate in the zone of carbonatization and very high secondary porosities in the underlying zone of decarbonatization.

6) Mesogenetic sandstone decarbonatization is generally controlled by the amount and type of organic matter in intercalated shales and their temperature-time exposure. The texture and composition of the non-carbonate fraction of the sandstones appears to have little influence on decarbonatization. Significantly, decarbonatization can create porosity in tight sandstones because of the presence of irreducible lamellar porosity.

7) Secondary porosity in any sandstone lithology resists reduction by chemical compaction much more than does primary sandstone porosity. However, the rates of reduction of both primary and secondary porosity by chemical compaction are influenced by temperature and mineralogical maturity of the sandstones. Therefore, mapping of the diagenetic maturity of sandstones is only of value if there is reasonable uniformity of mineralogical maturity.

8) The maximum depth of effective secondary porosity created by mesogenetic decarbonatization is more than double that of effective primary porosity in sandstones of high mineralogical stability. It is more than three times the

maximum depth of effective primary porosity in sandstones of low mineralogical stability.

9) The maturation of the hydrocarbon source rocks appears to be directly responsible for the generation of most secondary sandstone porosity. The majority of conventional crude oil and natural gas in clastic sequences has been generated in the source rocks after the intercalated sandstones had lost their primary porosity and after they had gained their secondary porosity. For this reason secondary sandstone porosity is a favorable habitat for crude oil and natural gas in clastic sequences.

10) Sandstone reservoirs with secondary porosity usually have pore and reservoir geometry and other petrophysical and reservoir characteristics that are strikingly different from those of equivalent sandstone reservoirs with primary porosity. Secondary porosity in sandstones poses special problems for exploration and production of energy resources, but it also offers specific opportunities for refining exploration and production technology in order to optimize the efficiency of the search for and the recovery of hydrocarbons.

ACKNOWLEDGMENTS

Research for this paper was conducted at the geological laboratory of Mobil Oil Canada, Ltd. and since 1976 also at the geological research laboratory of Petro-Canada. The authors would like to thank both companies for their generous support in every phase of the research and the preparation of these course notes.

Special thanks are due to Susan Peters and Brian Miller for typing and assembling these course notes.

REFERENCES

ALCOCK, F.G., and R.I. BENTEAU, 1976, Nipisi Field - a Middle Devonian clastic reservoir: in M.M. Lerand (ed.), The sedimentology of selected clastic oil and gas reservoirs in Alberta; Can. Soc. Petrol. Geol., 125 p.

CHOQUETTE, P.W. and L.C. PRAY, 1970, Geologic nomenclature and classification of porosity in sedimentary carbonates: Am. Assoc. Petrol. Geologists Bull., v. 54, p. 207-250.

FOX, J.E., LAMBERT, P.W., MAST, R.F., NUSS, N.W., and R.D. REIN, 1975, Porosity variation in the Tensleep and its equivalent the Weber Sandstone, Western Wyoming; a log and petrographic analysis: in D.W. Bolyard (ed.), Deep drilling frontiers of the central Rocky Mountains; Rocky Mountain Assoc. Geologists.

GALLOWAY, W.E., 1974, Deposition and diagenetic alteration of sandstone in northeast Pacific Arc-related basins; implications of greywacke genesis: Geol. Soc. America Bull., v. 85, p. 379-390.

HARTMAN, J.A., 1968, The Norphlet Sandstone, Pelahatchie Field, Rankin County, Mississippi: Gulf Coast Assoc. Geol. Soc., Trans., v. 18, p. 2-11.

HAYES, J.B., 1979, Sandstone diagenesis - the hole truth: in P.A. Scholle and P.R. Schluger (eds.), Aspects of diagenesis - symposia: Soc. Econ. Paleontologists and Mineralogists Spec. Pub. 26.

HEALD, M.T., and R.E. LARESE, 1973, The significance of solution of feldspar in porosity development: Jour. Sed. Petrology, v. 43, p. 458-460.

HRABER, S.V., and P.E. POTTER, 1969, Lower West Baden (Late Mississippian) sandstone body of Owen and Green Counties, Indiana: Am. Assoc. Petrol. Geologists Bull., v. 53, p. 2150-2160.

LINDQUIST, S., 1976, Leached porosity in overpressured sandstones - Frio Formation (Oligecene), South Texas (abs.): Am. Assoc. Petrol. Geologists Bull., v. 60, p. 1612.

KRYININE, P.D., 1941, Petrographic studies of variation in cementing material in the Oriskany sand: Proc. 10th Penna. Mineral Ind. Conf., Penn. State Coll., Bull. 33, p. 108-116.

PARKER, C.A., 1974, Geopressures and secondary porosity in the deep Jurassic of Mississippi: Gulf Coast Assoc. Geol. Soc., v. 24, p. 69-80.

PITTMAN, E.D., 1979, Porosity diagenesis and production capability of sandstone reservoirs: in P.A. Scholle and P.R. Schluger (eds.), Aspects of diagenesis - symposia: Soc. Econ. Paleontologists and Mineralogists Spec. Pub. 26.

PROSHLYAKOV, B.K., 1960, Reservoir rocks as a function of their depth and lithology: Geol. Nefti i Gaza, 4, No. 12, p. 24-29.

ROWSELL, D.M., and A.M.J. DeSWARDT, 1974, Secondary leaching porosity in Middle Ecca Sandstones: Trans. Geol. Soc. So. Africa, v. 77, p. 131-140.

SAVKEVIC, S.S., 1971, The possible connection between the leaching-induced porosity of rocks and the primary stage of petroleum generation (migration): IZV, AN SSR, Ser. Geol., 6, p. 70-77.

SCHMIDT, V., 1976, Secondary porosity in the Parsons Lake Sandstone (abs.): Geol. Assoc. Canada and Mineral. Assoc. Canada Ann. Mtg. Abstracts, v. 1, p. 50.

SCHMIDT, V., McDONALD, D.A., and R.L. PLATT, 1977, Pore geometry and reservoir aspects of secondary porosity in sandstones (abs.): Can. Soc. Petrol. Geol. and Can. Inst. Min. Ann. Mtg. Abstracts, p. 106.

SCHMIDT, V., and D.A. McDONALD, 1979(1), Texture and recognition of secondary porosity in sandstones: in P.A. Scholle and P.R. Schluger (eds.), Aspects of diagenesis - symposia: Soc. Econ. Paleontologists and Mineralogists Spec. Pub. 26.

SCHMIDT, V., and D.A. McDONALD, 1979(2), The role of secondary porosity in the course of sandstone diagenesis: in P.A. Scholle and P.R. Schluger (eds.), Aspects of diagenesis - symposia: Soc. Econ. Paleontologists and Mineralogists Spec. Pub. 26.

SCHOLLE, P.A.,1979, A color illustrated guide to constituents, textures, cements, and porosities of sandstones and associated rocks: Am. Assoc. Petroleum Geologists Mem. 28.

SEDIMENTATION SEMINAR, 1969, Bethel Sandstone (Mississippian) of western Kentucky and south-central Indiana, a submarine-channel fill: Kentucky Geol. Surv., Rept. Inv. 11, Ser. x, 24 p.

STANTON, G.D., and E.F. McBRIDE, 1976, Factors influencing porosity and permeability of Lower Wilcox (Eocene) Sandstone, Karnes County, Texas (abs.): Am. Assoc. Petrol. Geologists and Soc. Econ. Mineralogists and Paleontologists Ann. Mtg. Abstracts, v. 1, p. 119.

TISSOT, B, DURAND, B., ESPITALIE, J., and A. COMBAY, 1974, Influence of nature and diagenesis of organic matter in formation of petroleum,: Am. Assoc. Petrol. Geologists Bull., v. 58, p. 499-506.

WARDLAW, N.C., 1976, Pore geometry of carbonate rocks as revealed by pore casts and capillary pressure: Am. Assoc. Petrol. Geologists Bull., v. 60, p. 245-257.

-- Notes --